Introduction to Power Utility Communications

For a listing of recent titles in the
Artech House Power Engineering Series,
turn to the back of this book.

Introduction to Power Utility Communications

Dr. Harvey Lehpamer

ARTECH HOUSE
BOSTON | LONDON
artechhouse.com

Library of Congress Cataloging-in-Publication Data
A catalog record for this book is available from the U.S. Library of Congress.

British Library Cataloguing in Publication Data
A catalog record for this book is available from the British Library.

ISBN-13: 978-1-63081-006-1

Cover design by John Gomes

10 9 8 7 6 5 4 3 2 1

I dedicate this book to my dear wife, Monica,
for her endless patience and love

Contents

CHAPTER 7

Preface

Power utilities use communications systems for a variety of functions, from common voice and data applications, to SCADA and protective relaying, to security and maintenance. This book introduces the practical world of communications, emphasizing applications in the electric utility space.

The book is intended for those who need an introduction into evolving communications technologies as a prerequisite for the future *smart grid*—whether or not they specialize in communications engineering. Moreover, colleges and universities can adopt the book as a textbook for communications-related courses, including those targeted to power-track engineers and project managers.

This book is a useful practical resource for young engineers who are just entering the work force and facing the complexities of communications projects; instructors and consultants in the telecommunications field; and anyone else involved in real-life communications network buildout.

Our world is not standing still: New and improved technologies, updated standards, and new approaches that allow us to accomplish things faster and more efficiently are all pushing us to strive toward new goals. Accordingly, this book aims to make at least a small contribution toward expanding its readers' knowledge in communications systems design and deployment.

The book's extensive glossary defines many commonly used terms in the communications field and thus enables better comprehension of the material in this book.

For further details on some specific and more in-depth microwave radio, RFID, and fiber-optic topics, interested readers should consult one of my other three publications (*Transmission Systems Design for Wireless Networks*, *Mircrowave Transmission Networks—Planning, Design, and Deployment*, and *RFID Design Principles*) in the communications arena.

CHAPTER 1

Introduction to Communications Systems

1.1 Brief History of Communications

The transmission of signals over a distance for the purpose of communication began thousands of years ago with the use of smoke signals and drums in Africa, America, and parts of Asia and courier networks in Egypt, but it was not until the 1830s that electrical telecommunications[1] systems started to appear. It is interesting to note that the sources of today's communications network problems are very similar to those of many centuries ago: speed, reliability, and security.

Generally speaking, a communications network is a network of links and nodes arranged so that messages may be passed from one part of the network to another. Three types of media can be used in transmitting information in the telecommunications world: copper lines (twisted-pair and coaxial cables), optical fibers, and radio.

1.1.1 Alexander Graham Bell, Elisha Gray, and the Invention of the Telephone

In 1843, Samuel Morse invented the first long-distance electric telegraph[2] line, and Alexander Bain patented the first fax machine. The telegraph and telephone are both wire-based electrical systems, and Alexander Graham Bell's success with the telephone came as a direct result of his attempts to improve the telegraph.

In the 1870s, inventors Elisha Gray and Alexander Graham Bell independently designed devices that could transmit speech electrically (the telephone). Both men rushed their respective designs to the patent office within hours of each other; Bell patented his telephone first. Subsequently, Gray and Bell entered into a famous legal battle over the invention of the telephone, which Bell won in 1876.

Bell's unceasing and very broad scientific curiosity led to the invention of the photophone, to significant commercial improvements in Thomas Edison's phonograph, and to the development of his own flying machine just six years after the Wright brothers launched their plane at Kitty Hawk. As President James Garfield

1. The word telecommunication is adapted from a French word; it is a compound of the Greek prefix *tele*, meaning "far off," and *communication*, meaning "to transfer information."

2. 1844: First "long distance" telegraph connection between Baltimore and Washington.

lay dying from an assassin's bullet in 1881, Bell in a matter of days invented a metal detector in an unsuccessful attempt to locate the fatal bullet in the president's body. (Unfortunately, the president had one of the first metal beds of that era, and it interfered with the metal detector, so the bullet was never located.)

The photophone was similar to a contemporary telephone, except that it used modulated light as a means of wireless transmission while the telephone relied on modulated electricity carried over a conductive wire circuit. Of the 18 patents granted in Bell's name alone, and the 12 he shared with his collaborators, four were for the photophone, which Bell referred to as "the greatest invention he has ever made, greater than the telephone." It was an advanced form of telecommunications, but subject to atmospheric interference and impractical until the secure transport of light that would be offered by fiber-optic systems.

1.1.2 Nikola Tesla and the Invention of Radio

Although widely accepted as an inventor of radio, Guglielmo Marconi, an Italian engineer working in England, based his work on the patents and inventions of Nikola Tesla.[3] It was Tesla's original concept demonstrated in his famous lecture at the Franklin Institute in Philadelphia in 1893 that Marconi used for his radio transmission.

On November 8, 1898, Tesla patented a radio-controlled robot-boat. Tesla used this boat, which was controlled by radio waves, in the Electrical Exhibition in 1898 at Madison Square Garden.

Marconi received—jointly with the German physicist Karl Ferdinand Braun—the 1909 Nobel Prize in physics for their "contribution to the development of wireless telegraphy." Tesla[4] refused to share a Nobel Prize in Physics for the radio transmission with Marconi. Tesla's patents from 1900 were reversed in favor of Marconi in 1904 after large private investments (including Edison) were made toward Marconi's company. In 1943, the U.S. Supreme Court upheld Tesla's radio patent from 1900, and as it stands today, Tesla is an official inventor of radio[5] (Figure 1.1).

Tesla was, an inventor of (among other things) alternate current, used today in every household. Although he is often nearly forgotten, Tesla holds over 700 patents—and those are just the ones he remembered to patent.[6]

3. Nikola Tesla was born in 1856 in Smiljan, Croatia. Recently, the birthplace home of Nikola Tesla was renovated, restored, and turned into an interactive museum.
4. http://www.teslasociety.com/radio.htm
5. According to some cynics, the Supreme Court had awarded the patent to Tesla (a few months after his death) in order to avoid having to pay any royalties to the Marconi Corporation, which was suing the U.S. government for patents used during World War I.
6. In 1917, Tesla was awarded the Edison Medal, the most coveted electrical prize in the United States. Nikola Tesla's name has been honored with an International Unit of magnetic flux density called "tesla". The United States Postal Service honored Tesla with a commemorative stamp in 1983. Tesla was inducted into the Inventor's Hall of Fame in 1975.

Figure 1.1 Tesla and the radio device in a New York hotel.

1.1.3 Invention of Optical Communications

As mentioned earlier, the idea of using light to carry communication signals originated with Bell. However, this idea had to wait some 80 years for low-cost electronics and fiber manufacturing advancements for it to become useful in practical situations [1]. Jun-ichi Nishizawa, a Japanese scientist at Tohoku University, also proposed the use of optical fibers for communications in 1963. In addition, Nishizawa invented other technologies that contributed to the development of optical fiber communications, such as the graded-index optical fiber as a channel for transmitting light from semiconductor lasers.

The first working fiber-optical data transmission system was demonstrated by German physicist Manfred Börner at Telefunken Research Labs in Ulm in 1965; it was followed by the first patent application for this technology in 1966.

Charles K. Kao and George A. Hockham of British company Standard Telephones and Cables (STC) were the first to promote the idea that the attenuation in optical fibers could be reduced below 20 dB/km, making fibers a practical communications medium.[7] They proposed that the attenuation in fibers available at the time was caused by impurities that could be removed, rather than by fundamental physical effects such as scattering. They correctly and systematically theorized the light-loss properties for optical fiber, and pointed out the right material to use for such fibers: high-purity silica glass. This discovery earned Kao the Nobel Prize in Physics in 2009.

1.2 Networks and Data Transmission Media

In telecommunications, information can be transmitted between two locations using a signal that can be either analog or digital in nature. In today's telecommunications networks, digital transmission is used almost exclusively, by converting analog traffic, such as voice calls, to digital signals to facilitate long-distance transmission and switching. Many other analog signals from nature (e.g., temperature, pressure, and movement) are also converted to digital signals.

7. In 1966, Kao and Hockham published their seminal paper, "Dielectric-fiber Surface Waveguides for Optical Frequencies." Hockham had worked on theoretical aspects of optical fibers while Kao studied light losses. July 1966 is the date now regarded as the birthday of optical fiber communication.

1.2.1 Review of Logarithms and Decibels

In telecommunications, it is common to use logarithms[8] and logarithm-based units of measurement, whose basic properties we will now review.

The question is: why do we need logarithms? First, it is because humans hear things logarithmically (i.e., when a person says that something has doubled in loudness, the transmission power has actually increased by a power of 10). This same concept of the increase in power is applied to other communications systems. Second, we need logarithms because boosts or drops in signals are additive or subtractive (i.e., easier calculations); last, as the numbers of interest range from very small to very large, a logarithmic scale allows us to view them in comparison to each other.

Telecommunications most commonly uses the base 10 and sometimes uses base 2 (Napierian) and base of natural logarithms (e).

Communications uses decibels to compare one power (or voltage level) to another. The abbreviation for decibel, dB, was derived from Bell's last name.

$$\text{Gain or Loss in decibels} = 10\log\left(\frac{P_2}{P_1}\right) \tag{1.1}$$

where P_1 = input power and P_2 = output power.

If $P_2 > P_1$, the communications system has a gain, while in the case of $P_2 < P_1$ the communications system has experienced a loss. (See Table 1.1.)

The relationship of decibel loss and gain to power is known as the "3–10 rule." In calculating the power output of a communication system, the following calculations can be used as a quick reference:[9]

- For every 3 dB of gain, the power is doubled. For every 3 dB of loss, the power is halved.

Table 1.1 Decibel Gain and Loss

Decibel Gain (Power)	*Factor*	*Decibel Loss (Power)*	*Factor*
0	1	0	1
1	1.25	–1	0.8
3	2	–3	0.5
6	4	–6	0.25
10	10	–10	0.1
20	100	–20	0.01
30	1,000	–30	0.001

8. Napier coined the Latin word logarithmus which derives its meaning from two Greek words: logos meaning a principle relationship between numbers or ratio, and arithmos meaning number.

9. For more on logarithms and logarithmic and exponential equations, see Appendix A. For more information on decibels and decibels per milliwatt see Appendix C.

- For every 10 dB of gain, the power is increased 10 times. For every 10 dB of loss, the power is one tenth.

Power in communication systems is commonly expressed in a decibel over 1 milliwatt (dBm), or in other words, dBm is a power relative to 1 mW.

$$\text{Power} = 10\log_{10}\frac{\text{Power (mW)}}{1\text{mW}} \quad [\text{dBm}] \tag{1.2}$$

For example, a 100-mW signal is equivalent to a 20-dBm signal.

In addition, radio systems sometimes use a power relative to 1W or dBW.

$$\text{Power} = 10\log_{10}\frac{\text{Power (W)}}{1\text{W}} \quad [\text{dBW}] \tag{1.3}$$

For example, a 100-mW signal is equivalent to a –10-dBW signal.

Decibel relative to one milliwatt can be converted into decibel relative to one watt units by subtracting 30 dB.

$$\text{dBW} = \text{dBm} - 30 \tag{1.4}$$

so, for example,

$$+30\text{ dBm} = 0\text{ dBW}$$
$$-30\text{ dBW} = 0\text{ dBm}$$

1.2.2 Early Days of Electronics

Of the many major inventions that took place during the 20th century, some were enablers for today's communication technology explosion. Examples of important inventions in electronics (see Figure 1.2) that were crucial for the development and miniaturization of electronic devices are listed as follows:

- Vacuum tube (Lee de Forest, 1906);
- Transistor (Shockley, Bardeen, Brattain, 1947);
- Integrated circuit (IC) (Texas Instruments, 1958).

In 1883, Thomas Alva Edison discovered that electrons would flow from one metal conductor to another through a vacuum.[10] In 1904, John Fleming applied

10. This discovery of conduction became known as the Edison effect.

Figure 1.2 Fifty years of active electronic components.

the Edison effect in inventing a two-element electron tube called a diode, and Lee De Forest followed in 1906 with the three-element tube, the triode. These vacuum tubes made it possible to manipulate electrical energy so that it could be amplified and/or transmitted.

Early radio consisted of either radio telegraphy (the transmission of Morse code signals) or radio telephony (voice messages). Both relied on the triode and made rapid advances thanks to armed forces communications during World War I. In 1918, Edwin Armstrong invented the superheterodyne receiver, which could select among radio signals or stations and receive distant signals, and as a direct result, radio broadcasting grew fast in the 1920s. Only amplitude modulation (AM) had been used from 1920 to 1935, when Armstrong also invented wide-band frequency modulation (FM). After the war, electron tubes were used to develop the first computers,[11] but they were impractical because of the sizes of the electronic components.

The transistor functions like the vacuum tube, but it is tiny by comparison, weighs less, consumes less power, is much more reliable, and is cheaper to manufacture with its combination of metal contacts and semiconductor materials. Engineers from Bell Laboratories received a Nobel Prize for the creation of transistors.[12]

ICs[13] combined transistors and diodes (active devices) and capacitors and resistors (passive devices) on a planar unit or chip. By the 1960s, ICs were in full production at a number of firms, and designs of electronic equipment changed rapidly to take advantage of the technology. Digital ICs with bipolar transistors were made first, but analog ICs, large-scale integration (LSI), and very-large-scale integration (VLSI), consisting of thousands of circuits on a single chip, followed by the mid 1970s. Microcomputers, medical equipment, video cameras, digital com-

11. Electronic Numerical Integrator and Computer (ENIAC) was the first electronic general-purpose computer. Designed in 1946, ENIAC contained 17,468 vacuum tubes; 7,200 crystal diodes; 1,500 relays; 70,000 resistors; 10,000 capacitors; and around 5 million hand-soldered joints.

12. The Nobel Prize in Physics 1956 was awarded jointly to William Bradford Shockley, John Bardeen and Walter Houser Brattain for their research on semiconductors and their discovery of the transistor effect.

13. The concept of the integrated circuit (IC) was proposed in 1952 by Geoffrey W. A. Dummer, a British electronics expert with the Royal Radar Establishment. In 1958, Jack S. Kilby of Texas Instruments and Robert Noyce of Fairchild Semiconductor, working independently, created the integrated circuit, a composite semiconductor block in which transistor, resistor, condenser, and other electrical components are manufactured together as one unit.

munications, and satellites enabling human space exploration are just some of the advances that ICs made possible [2].

Many early major civilian technology and electronics inventions coincided with other, at first glance, seemingly unrelated events. In other words, they were indirect or even unintentional results of war efforts or space exploration. Some examples are vacuum tubes and radio receivers in the early 1900s (WW1); radar and microwave communications (and interestingly, microwave ovens[14]) in the 1930s and 1940s (WW2); and transistors in the late 1940s and ICs in the 1950s and 1960s (space exploration).

1.2.3 Introduction to a Communications System

1.2.3.1 Communications System Building Blocks

Communication is the conveyance of information (message) from a source to a destination. (See Figure 1.3.)

There are five main building blocks of communications systems. Source is the originator of the message, whether it is a person or machine. Transmitter is the equipment that modifies the message (either data or voice) into the form required for transmission. The communications channel can be defined as a pathway over which information can be conveyed from the source to the destination; transmission media may be physical, like a copper wire or fiber-optic cable, or atmospheric, like radio waves. The receiver is the device that captures the message from the communications channel and converts it into a form that the person or machine at the destination can understand. The destination is the person or machine to whom the message is directed.

Common questions like, "What are the design parameters of a communication system?" and, "Is there a limit to how fast one can communicate?" are detailed in Chapter 2. This chapter discusses the main communications system parameters: bandwidth, capacity, signal-to-noise ratio (SNR) (analog systems), bit error ratio (BER) (digital systems), end-to-end transmission delay, and maximum link distance (signal attenuation and signal distortion).

1.2.3.2 Sources, Destinations, and Flow of Information

Flow of information between nodes can be simplex, half-duplex, or full duplex (Figure 1.4), described as follows.

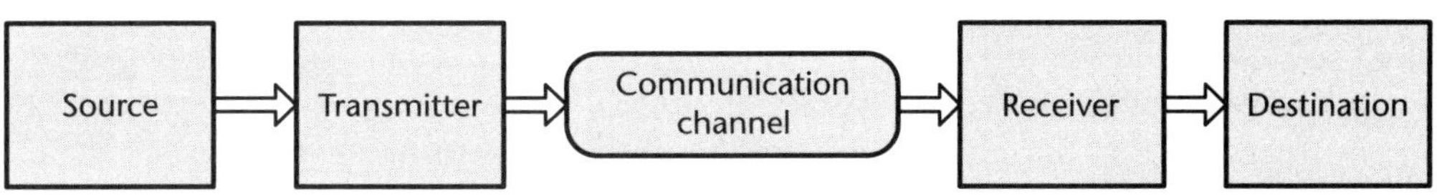

Figure 1.3 Communications system building blocks.

---

14. In 1947 Raytheon patented the dielectric heating device, naming it the Radarange, resulting in the first commercially available microwave oven. It was an 800-pound device priced at thousands of dollars and mainly used for commercial purposes.

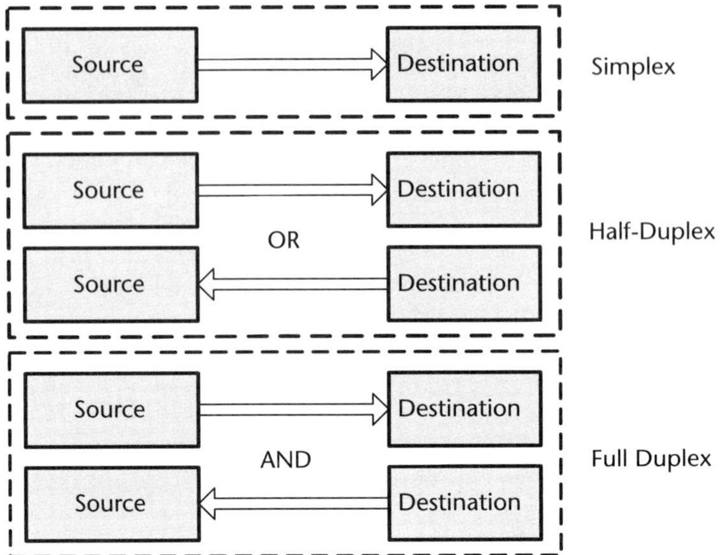

Figure 1.4 Flow of information.

- Simplex communication: Messages flow in only one direction, from source to destination. For example, in TV and radio broadcasting, information flows only from the transmitter site to multiple receivers. Here, connectionless protocols are used, as no acknowledgement or return traffic is possible over a simplex circuit. Satellite communication is also an example of simplex communication. [3]
- Half-duplex communication: Messages travel in both directions between the source and the destination, but in only one direction at a time. Half-duplex systems are usually used to conserve bandwidth, since they need only a single communications channel, which is shared alternately between the two directions. For example, a walkie-talkie requires only a single frequency for bidirectional communication, while a cell phone, which is a full-duplex device, requires two frequencies to carry the two simultaneous voice channels, one in each direction.
- Full-duplex communication: Messages can travel over the communications channel in both directions simultaneously. Where channel access methods are used in point-to-multipoint networks (such as cellular networks) for dividing forward and reverse communication channels on the same physical communications medium, they are known as duplexing methods, such as time-division duplexing (TDD) and frequency-division duplexing (FDD).

Although the required information for transmission originates from a single source, there may be more than one destination (more than one receiver), depending upon how many receive stations are linked to the channel and how much energy the transmitted signal possesses. (See Figure 1.5.)

In one-to-one communication, a single source sends information to a single destination. In one-to-many communication, a single source simultaneously sends

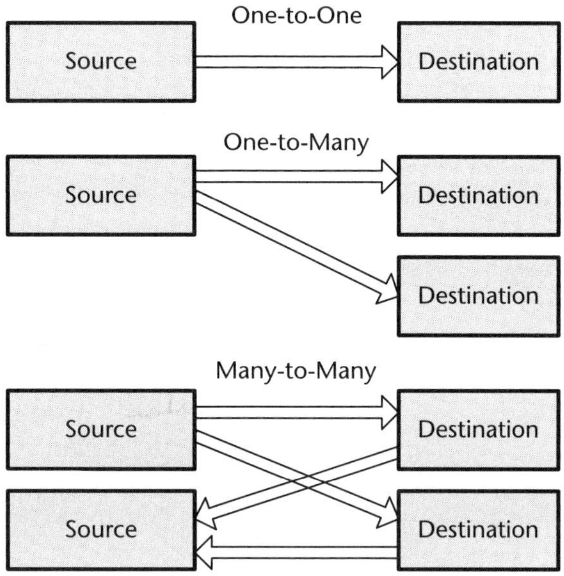

Figure 1.5 Information sources and destinations.

information to multiple destinations, and many-to-many communication occurs when many sources issue messages to many destinations.

1.2.4 Local Area Networks (LANs) and Wide Area Networks (WANs)

One way to categorize the different types of communications network designs is by their scope or scale. For historical reasons, the networking industry refers to almost every type of configuration as some kind of "area network" [4]. Common area network types are described as follows:

- LANs: Serve users within a confined geographical area (usually within a mile). In the power utility world, this would be an equivalent of the substation environment. Short links of fibers and/or copper cables and radio systems are typically found here.
- Wireless LANs (WLANs): Similar to LANs, but they use radio connectivity.
- WANs: Cover a wide geographical area, such as a state or a country, and connects these areas via hundreds of miles of fiber-optic systems (over transmission lines) and microwave systems.
- Metropolitan area networks (MANs): Cover a geographic area the size of a city or suburb and connectivity of their electric substations via fiber-optic or microwave systems (including leased lines from telecom carriers). Customers are also part of MANs as their homes are part of the communication network as well.
- Other area networks: Examples include storage area networks (SANs), campus area networks (CANs), personal area networks (PANs), field area networks (FANs), home area networks (HANs), neighborhood area networks (NANs) and body area networks (BANs).

Equipment used for interconnecting LANs and WANs are hubs, bridges, switches, and routers (Figure 1.6), described as follows.

- Hubs: Interconnecting two or more workstations into a LAN, hubs are simple interconnecting devices that require no overhead to operate. When a workstation transmits to a hub, the hub immediately resends the data frame out on all connecting links. A hub can be managed or unmanaged; a managed hub possesses enough processing power that it can be managed from a remote location.
- Bridges: They can be used to connect two similar LANs, such as two carrier sense multiple access with collision detection (CSMA/CD) LANs. A bridge can also be used to connect two similar LANs, such as a CSMA/CD LAN and a token ring LAN. The bridge examines the destination address in a frame and either forwards this frame onto the next LAN or does not. The bridge examines the source address in a frame and places this address in a routing table to be used for future routing decisions. Unlike routers, bridges are protocol-independent. They simply forward packets without analyzing and rerouting messages.
- Switches: A combination of a hub and a bridge, switches can interconnect two or more workstations, but like bridges, they observe traffic flow and learn. When a frame arrives at a switch, the switch examines the destination address and forwards the frame out the one necessary connection. Switches are considered layer-two devices, using media access control (MAC) addresses to forward frames to their proper destination.
- Routers: Layer-three devices, routers are much more complex and used to route packets between networks based on their Internet protocol (IP) address. Routers provide the following services: route discovery; selection of

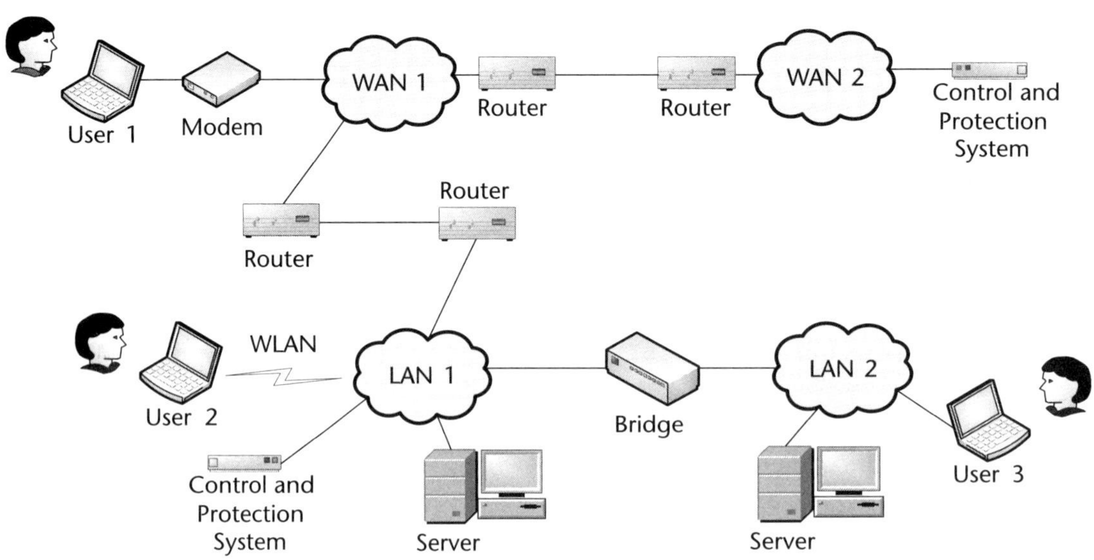

Figure 1.6 LANs and WANs.

the best route to a destination; adaptation to changes in the network; translation from one technology to another, such as Ethernet[15] to token ring; packet filtering based on IP address, protocol, or UDP/TCP port number; and connection to a WAN. Routers often incorporate firewall functionality that provides a mechanism to filter or redirect packets that do not fit security or access policy requirements.

1.2.5 Transmission Media

The main objectives of the transmission network (Figure 1.7) are to connect all the points of interest, satisfy the capacity demands, and provide reliable service using transmission media. The terms transmission and transport are used interchangeably in this text, as both are currently in use; the former is preferred in Europe, but the latter is more commonly used in North America. Sometimes transmission refers only to the physical media, while transport can include other open systems interconnection (OSI) model layers of the data transfer.

The term *backhaul* is commonly used for the WAN type (higher bandwidth and longer distances) of networks. The term *backbone* is used usually for high-capacity communications systems in ring or redundant topology.

Wireline systems are represented by copper wires and optical fibers and power line carriers in electric power utilities. There are a wide variety of wireless systems,

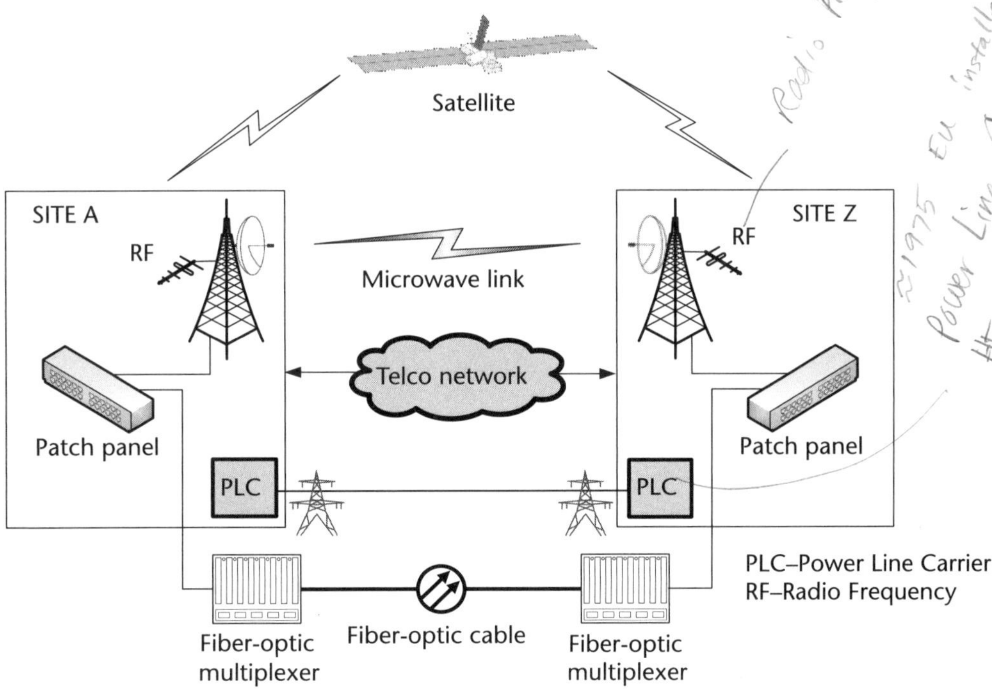

Figure 1.7 Transmission networks.

15. Ethernet is the most widely used wired local area network. Ethernet uses carrier sense multiple access (CSMA) to allow computers to share a network and operates at 10, 100, or 1,000 Mbps, depending on the physical layer used.

including radio systems in lower frequency bands and microwave- and millimeter-wave frequency bands, short-range and long-range, terrestrial and satellite, and free-space laser communications. Circuits leased from the telecom companies, although considered copper lines, may be going over telco-owned fiber-optic and/or microwave links as well [5].

1.2.5.1 Copper Lines

Years ago, copper wire was the only means of transporting information. Technically known as unshielded twisted pair (UTP), it consists of a large number of pairs of copper wire of varying size within a cable. The cable did not have a shield, so the signal (primarily the high-frequency part of the signal) was able to leak out. In addition, the twisting on the copper pair was very casual, designed as much to identify which wires belonged to a pair as to handle transmission problems. Even with these limitations, it was quite satisfactory for use in voice communications.

Coaxial cable technologies were primarily developed for the cable TV industry. In the last few years, this technology has been extended to provide Internet services to residences. The high capacity of coaxial cable allows it to support multiple TV channels, and this capacity can also be used for high-speed Internet access. Coaxial cable is widely used to connect radio transmitters and receivers to indoor or outdoor antennas.

Ideally, regardless of the copper cable type, the signal at the receiver should be the same as at the transmitter. Unfortunately, this is not true in real cables, and signal quality degrades for several reasons, including attenuation, crosstalk, and impedance mismatches. All signals degrade when transmitted over a distance through any medium (including optical fibers).

Attenuation refers to decrease in signal strength, measured in decibels per unit length (usually decibels per kilometer), and in copper cables, it increases with frequency. Attenuation in copper cables is primarily dependent on the resistance of the conductors, while in optical fibers, the transmitted light is attenuated because it is absorbed by the natural properties of the glass fiber.

Figure 1.8 illustrates what happens with a digital signal with very fast rising edges that contain high frequencies. In a transmission channel, the higher frequencies are attenuated more than the lower frequencies, resulting in signal attenuation and distortion that increase with the distance. On a long wireline system, amplifiers or repeaters must be used at regular intervals along the channel to restore the signal to acceptable levels. While amplifier increases a level of useful signal, it also amplifies the noise and distortion; repeaters (also called regenerators) not only increase

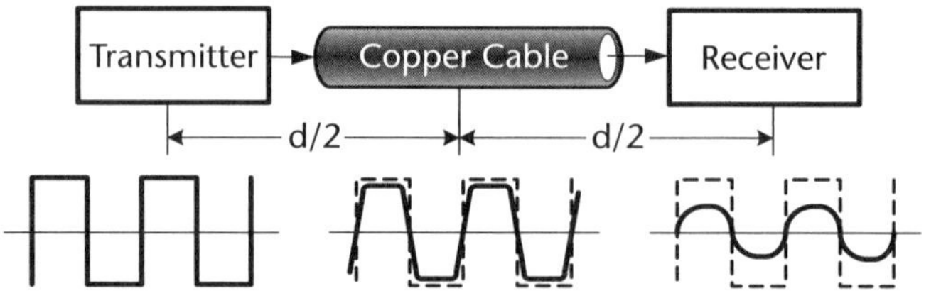

Figure 1.8 Signal attenuation and distortion.

the signal level but also completely restore the signal to what it was when generated in the transmitter.

Crosstalk is interference in the form of an electric signal from an adjacent cable or circuit. A commonly used measure of this interference in twisted-pair cable is near-end crosstalk (NEXT), shown in decibels. The higher the decibel value, the less the crosstalk, and therefore the better the cable. Shielding of the cable is the most common way to decrease the effects of crosstalk. Crosstalk does not exist in fiber-optic cables.

The characteristic impedance (Z_0) of a cable is the value of impedance that characterizes that cable. The characteristic impedance is the input impedance of the cable seen when the cable is terminated with the characteristic impedance, as shown in Figure 1.9. Such a cable then appears electrically to be infinitely long and has no signal reflections from the termination load (Z_L). If one cable is connected to another cable (or another device, like an antenna) of differing characteristic impedance, then signals are reflected at their interface. These reflections caused by impedance mismatch have to be completely avoided (in an ideal world) or at least minimized (in the real word) as much as possible.

Chapter 7 details impedance mismatch and return loss measurements.

1.2.5.2 Optical Fibers

Optical fibers constitute the second transmission medium, and this is unquestionably the high-bandwidth transmission medium of choice today. Transmission speeds of as high as 10 Gbps have become commonplace in the industry.

There is a wide range of types of glass available, and characteristics vary depending on chemical composition. Over the past few decades the transmission properties of glass have been improved considerably. In the 1970s, the attenuation of a silicon fiber was close to 20 dB/km. By the 1980s, research had improved this to 1 dB/km, and in the 1990s, average attenuation reached 0.2 dB/km, which is still the case today. Most of the attenuation in fiber is caused by light being scattered by minute variations (less than one-tenth of the wavelength) in the density or composition of the glass. This is called Rayleigh scattering.[16]

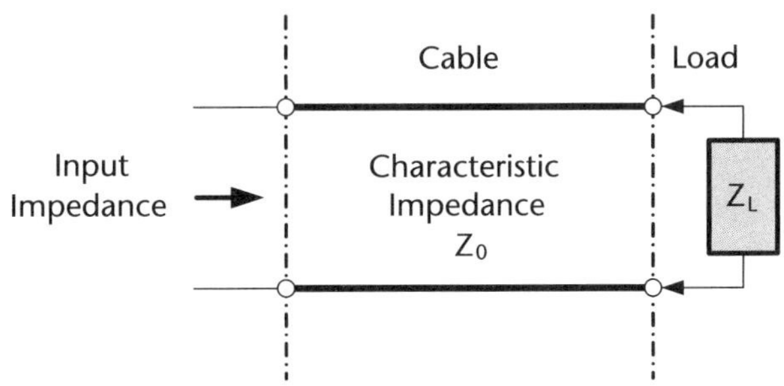

Figure 1.9 Characteristic impedance of the cable.

16. Rayleigh scattering is also the reason that the sky is blue and that sunsets are red. In fiber, Rayleigh scattering is inversely proportional to the fourth power of the wavelength, accounting for most of the enormous difference in attenuation of light at 850 nm wavelength from that at 1,550 nm. Unfortunately, we can't do a lot about Rayleigh scattering by improving fiber manufacturing techniques.

Fiber-optic cables can be placed in ducts, buried in the ground, suspended in the air between poles, and installed as part of the ground wire on the high-voltage transmission towers—optical ground wire (OPGW)—among other options. Of course, laying fiber, on a per-mile basis, still costs somewhat more than laying copper, but on a per-circuit basis there is no doubt that fiber is more cost-effective.

The huge capacity of fiber certainly makes for more efficient communications; however, placing so much traffic on a single strand, for point-to-point communications, makes for greater vulnerability. Most disruptions in fiber-optic networks are a result of a physical cut of a fiber run (called backhoe fade),[17] and the ring configuration is the protection solution used most often in fiber-optic networks.

Two most commonly used types of optical fibers are multimode fibers (for short distances of up to few miles) and singlemode fibers (for long distances).

1.2.5.3 Radio and Other Wireless Systems

Wireless communications can take several forms, including microwave (point-to-point or point-to-multipoint), synchronous satellites, low Earth orbit satellites (LEOs), cellular, personal communications service (PCS), and free-space optic communications. To understand wireless technology, a basic understanding of the radio frequency (RF) spectrum is required.

The RF spectrum is a part of the electromagnetic spectrum in which a variety of commonly used devices (including television, AM and FM radios, microwave radios, cell phones, and pagers) operate. The electromagnetic spectrum has been used for communications for over 100 years, and it comprises an infinite number of frequencies, from AM radio at 1 MHz to the cellular/PCS band at 2 GHz and microwave- and millimeter-wave frequency bands.[18]

Frequencies are measured in cycles per second, or hertz, which are inversely related to wavelength. At low frequencies wavelengths are long, while at higher frequencies wavelengths are very short. Given an equal power level, the longer the wavelength, the greater the distance the signal can travel. Where low-frequency signals (such as AM radio) can be transmitted for hundreds of miles, high-frequency signals (such as infrared) can travel only a few feet.

For years, microwave radio transmissions have been used in the telecommunications industry for the transport of point-to-point data where information transmissions occur through carrier signals. Microwave carrier signals are typically relatively short in wavelength and can transmit information using various modulation methods.

The fixed-satellite service basically involves four frequency bands: 4/6 GHz, 7/8 GHz (for military systems), 11/14 GHz, and 20/30 GHz. Although there are numerous bands above 30 GHz allocated to the fixed-satellite service, only one is presently being used. Microwave frequencies and "stationary" satellites allow the use of high-gain, directional antennas, much like the fixed service, reducing power requirements for the satellite transmitters. The fixed-satellite service includes

17. Backhoe fade is a term coined by the telecommunications industry, referring to the accidental severing of a cable by a backhoe or similar construction activity. The term comes from the sudden and initially inexplicable loss of signal ("fading") experienced when a cable is accidentally dug up and damaged.

18. 1 GHz = 1,000 MHz = 1,000,000 Hz

international, domestic, and military systems and, although they often carry the same type of traffic, each group has its own set of users. International and domestic systems operate at 4/6 GHz and 11/14 GHz, while military systems use 7/8 GHz and frequencies near 20 GHz and 45 GHz.

Another one of the wireless telecommunications technologies is the LEO satellite system, comprised of satellites that communicate directly with handheld telephones on Earth. Because these satellites are relatively low (less than 900 miles above), they move across the sky quite rapidly. They operate similar to a terrestrial cellular system, except that in this case it is the cell site that is moving rather than the subscriber.

Free-space laser communications[19] systems are wireless, point-to-point connections through the atmosphere that employ the optical part of the frequency spectrum. Therefore, they cannot be categorized as either wireless or wireline systems in a classical sense. They work only under clear line-of-sight (LoS) conditions between each unit, eliminating the need for securing rights of way, buried cable installations, and government licensing. Free-space laser communications systems can be quickly deployed, since they are small and do not need any radio interference studies.

Although rain can cause attenuation up to 40 dB/km and snow even to 100 dB/km, fog is, by far, the largest problem. In extremely heavy fog, signal attenuation as high as 300 dB/km (500 dB/mi) has been reported.

1.2.5.4 FCC Licensing

The RF spectrum is subject to regulation and licensing by the Federal Communications Commission (FCC) in the United States,[20] Industry Canada in Canada, Cofetel in Mexico, and globally via the International Telecommunications Union (ITU).[21] In the United States, the National Telecommunications and Information Administration (NTIA) manages the federal government's use of spectrum. Countries that are members of the ITU generally follow the ITU spectrum allocation.

Within the RF spectrum, not all frequencies are subject to licensing requirements, and license-exempt bands include the industrial, scientific, and medical (ISM) band (the most widely used license-exempt frequency band) and the unlicensed national information infrastructure (UNII) band.

The frequencies that are used for radio communications have slowly moved upward from lower to higher frequencies (shorter wavelengths). Back in the early days of radio, it was easier to generate carrier frequencies of sufficient power at the lower side of the frequency spectrum. With the advancement of new techniques, it

19. This type of optical communication, also known as free-space optical (FSO), has emerged as a commercially viable alternative to RF and millimeter-wave wireless systems.

20. The August 2011 edition of the United States Frequency Allocations can be found at: http://www.ntia.doc.gov/page/2011/united-states-frequency-allocation-chart (accessed April 29, 2015).

21. Publications of the ITU are referred to as recommendations because the ITU has no enforcement authority over the member nations that use them. Its strongest influence is the ability to influence its member telecommunications authorities to use the standards because it makes sense to do so on a global basis. The recommendations are published every four years.

has become to possible to develop new components that work at higher and higher frequencies.

1.2.6 Security and Encryption

Digital information is often the target of computer hackers, international spies, and criminals. Encryption is the intentional scrambling or masking of digital data to protect it from compromise. In order to protect the information, in 1977 the National Security Agency (NSA) and the National Bureau of Standards (NBS) adopted the data encryption standard (DES) to protect sensitive, unclassified, nonmilitary digital information from unauthorized access.

The DES algorithm is the most widely used encryption algorithm in the world. For many years, and among many people, "secret code-making" and DES have been synonymous. DES works by encrypting groups of 64 message bits, which is the same as 16 hexadecimal numbers. To do the encryption, DES uses "keys," which are also apparently 16 hexadecimal numbers long or apparently 64 bits long. However, every eighth key bit is ignored in the DES algorithm, so that the effective key size is 56 bits. In any case, 64 bits (16 hexadecimal digits) is the round number upon which DES is organized.

In cryptography, triple-DES is a block cipher formed from the DES cipher by using it three times. Given a plaintext message, the first key is used to DES-encrypt the message. The second key is used to DES-decrypt the encrypted message and, since the second key is not the right key, this decryption just scrambles the data further. The twice-scrambled message is then encrypted again with the third key to yield the final ciphertext.

The *advanced encryption standard* (AES) is a symmetric key encryption technique that has replaced the commonly used DES. AES offers markedly higher security margins, a larger block size, and potentially longer keys. One large-scale exception is within the electronic payments industry, which still uses triple-DES extensively and continues to develop and promulgate standards based upon it, guaranteeing that triple-DES will remain an active cryptographic standard in the future. By design, DES, and therefore triple-DES, suffer from slow performance in software; on modern processors, AES tends to be around six times faster.

AES[22] was the result of a worldwide call for submissions of encryption algorithms issued by the U.S. government's National Institute of Standards and Technology (NIST) in 1997 and completed in 2000 [6]. AES provides strong encryption and was selected by NIST as a federal information processing standard in 2001. The AES algorithm uses one of three cipher key strengths: a 128-, 192-, or 256-bit encryption key (password). Each encryption key size causes the algorithm to behave slightly differently, so the increasing key sizes not only offer a larger number of bits with which you can scramble the data, but also increase the complexity of the cipher algorithm. AES is fast in both software and hardware, is relatively easy to implement, and requires little memory.

22. The winning algorithm, Rijndael, was developed by two Belgian cryptologists, Vincent Rijmen and Joan Daemen. In 2003 the U.S. government announced that AES was secure enough to protect classified information up to the top secret level, which is the highest security level and defined as information that would cause "exceptionally grave damage" to national security if compromised.

The U.S. government has approved triple DES and the federal information processing standards (FIPS PUB197) to coexist with AES through the year 2030, thus allowing for a gradual transition to AES [7].

1.3 Electrical Grid and Power Utility Communications

1.3.1 An Overview

Power utilities currently use communications systems for providing important operational, business, and administrative applications. Power utilities use communications channels for a variety of functions, from common voice and data applications, to supervisory control and data acquisition (SCADA) and protective relaying (teleprotection), to security and maintenance.

It is common for power utilities to use technologies such as substation automation systems, surveillance video, telemetering, and on-line condition monitoring systems to help improve reliability and availability of power supply and to improve operational efficiency. Fault diagnostics and on-line monitoring tools such as fault recorders, fault locators, and condition monitoring systems are installed to assist operators and maintenance staff in determining the integrity, location and probable causes of faults quickly. Information is quickly sent to the operational and maintenance staff to minimize supply restoration time—or to avoid potential catastrophes.

Early utility monitoring and control systems used telephone technology (leased telephone lines) operating at 300 bps. Today, leased telephone lines (from local telephone companies) are still quite common; they operate at 1,200; 4,800; or 9,600 bps or higher.[23]

Early on, for communications to very remote hydroelectric power plants, power-line carrier systems (PLC) between high-voltage substations were used. Over the years, the majority of the PLC systems were replaced with microwave radio links and fiber-optic systems. Fiber-optic cable is currently being used both within substations and as WANs. OPGW is a fiber-optic solution used exclusively in electric utility space.

In the 1990s, license-exempt 900-MHz mesh radio systems were installed and added to the communications network options (but rarely for protection). Although the trend is to add higher speeds with more throughput capacity, many power utilities still operate with 1,200-bps leased lines.

Information and communications technology infrastructures, which are key components of grid modernization capabilities, are critical to communications with field resources during restoration efforts. Primary and backup communication paths to key control and monitoring points on the grid could be provided in multiple forms (e.g., private and public networks paired together).

Satellite communications are rarely used—generally only for sparsely populated (remote) large geographic areas and for systems that do not care about the high latency. Typically, this is a last option to be considered when it comes to utility

23. Most likely, by 2020 telecom companies in the United States will not be offering any circuit below T1 (1.544 Mbps) any longer..

communications due to the cost and high latency. They can also be used for voice communications in emergency situations.

There is not a single solution or a technology for the *smart grid*. Utilities use and will continue to use a wide variety of communications networks as they evolve the basic power grid into a smart grid. Security, compatibility, and interoperability of different systems are, and will remain, major challenges in the future.

1.3.2 SCADA

1.3.2.1 A Brief History of SCADA Systems

Some of the earliest SCADA systems were installed in the 1920s; subsequently, in the 1930s, individual utilities started interconnecting to interchange electricity to reduce operating costs. Analog computers were developed to control generation much more closely in order to monitor and control generator output, tie-line power flows, and frequency.

By the 1950s, analog computers were enhanced to schedule generation for each generator to provide the lowest cost of generation. These functions were called economic dispatch (ED) and automatic generation control (AGC), and the systems were called energy management systems (EMSs). In the beginning, these EMS[24] functions were supported by off-line manual calculations; in the late 1960s, digital computers and software replaced the analog EMS and manual calculations, and in the 1980s and 1990s, open standard operating systems were developed that supported real-time applications.

In the beginning of monitoring and control systems, the interface between the power system and the control system was in a remote location—and known as a remote terminal unit (RTU). It consisted of a cabinet or panel of terminals for the instrumentation and control wires, which connected it to the power system. The RTU remained basically the same until the mid 1970s when rugged microprocessors that could withstand the substation environment became available; in other words, they were substation-hardened.[25] The new and enhanced capabilities included time-keeping, more complex and powerful protocols, individual point numbering, local logging and time tagging of events, higher communications speeds, and multiple communications ports.

In the 1980s, microprocessors were applied to protective relays, meters, and various controllers and other devices that were usually equipped with a communications port. As these new and more powerful devices were deployed, utilities and manufacturers realized that substation design and complexity could be greatly simplified by interfacing these devices directly into the RTU. As the application of these devices grew, the IEEE Power and Energy Society (PES) substations committee determined that a need existed for a unique name to identify them, and so it

24. EMS is a system of computer-aided tools used by operators of electric utility grids to monitor, control, and optimize the performance of the generation and/or transmission system.

25. Hardening substation equipment means making it less vulnerable to the full range of threats to equipment integrity and system reliability, either natural or man-made. These include switching transients, as well as EMI from lightning strikes, electrostatic discharges, and even interference from work crew radios. Environmental conditions such as extreme temperatures and the emission of potentially disturbing surges or interfering noise that could harm other devices inside or outside the substation also need to be considered.

started calling them intelligent electronic devices (IEDs). In fact, today, any device with a microprocessor and a communications port is called an IED.

Today, visualization is one of the core elements in an EMS, and it is essential to manage large volumes of data. It acts as the interface between the EMS applications and the system operators and converts data and information into geometric or graphic representations. It provides an intuitive way to express the large quantity of complex data so that control engineers can respond to system emergencies and contingencies in a very short time. Visualization may include two-dimensional (2-D) presentation, three-dimensional (3-D) presentation, virtual reality, and animation [8].

1.3.2.2 Modern SCADA Systems

SCADA systems are specialized computer networks and devices that monitor and control key processes involved in the management of machinery, equipment, and facilities. (See Figure 1.10.)

Measurements taken from a variety of sensors (e.g., temperature, pressure, flow, and synchrophasors (temperature sensors, pressure sensors, flow sensors, sychrophasors, etc.) are used to make decisions, such as opening a valve and releasing water from a tank when it fills up, or initiating an emergency shutdown of a nuclear power plant.

Distribution automation (DA) and computerized remote control of medium-voltage (MV) substations and power grids help electric utilities to achieve higher reliability of supply and to reduce operating and maintenance costs. Today,

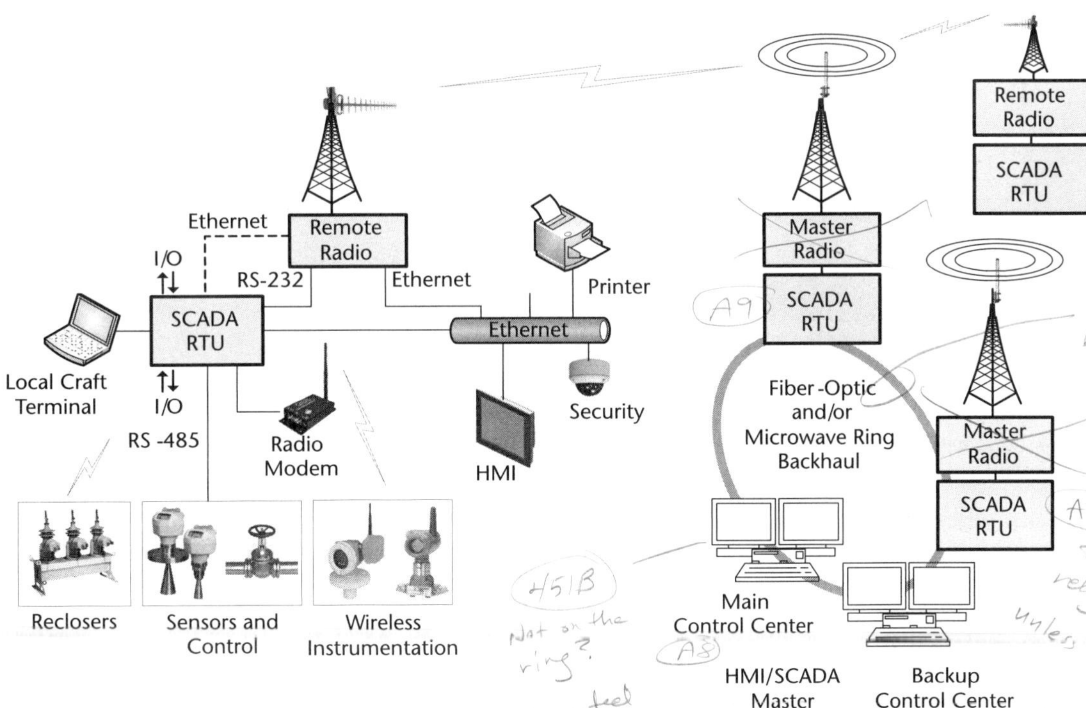

Figure 1.10 SCADA system block diagram.

sectionalizing equipment, switchgear, reclosers[26] and capacitor banks, and wind farm power generation connected to the MV power grid are all designed for remote operation using, in most cases, some type of radio communication.

A traditional SCADA implementation uses specific software and a personal computer to collect data from remote stations and display it in various forms for the user via the human machine interface (HMI). SCADA devices communicate with the control system using specialized protocols via the following interfaces:

- HMIs: They allow an operator to view and react to process status and events.
- Supervisory systems: Computers that monitor and send commands to control devices and processes.
- RTUs: They convert signals from process sensors to digital data and send them to the supervisory system.
- Communications infrastructure: It connects RTUs to the supervisory system by such means as PLC, leased lines, and RF modems.

Oil and gas operations require remote control for production wellheads, along long pipelines and at valves located at difficult-to-access sites. This industry must maintain the strictest safety measures to prevent and detect leakages or fire at remote installations. Thus, each site must be supervised via RTUs with communications to one or more control centers. RTUs in gas installations calculate gas flow according to AGA-3,[27] AGA-7, AGA-8 and other standard formulas, while RTUs along oil pipelines perform analog pressure control using proportional-integral-derivative (PID) loops.

Intrinsically safe[28] connection to RTUs from field sensors installed in restricted zones can be provided by adding barrier devices to the discrete input and output signals and EIA-232 serial cables.

Aside from power utilities, SCADA systems using RTUs are utilized for many other purposes, including water and wastewater industry applications, public safety and fire station alerting, natural disaster warning systems, and large industrial and manufacturing facilities.

1.3.3 Wide Area Measurement Systems (WAMSs) and Synchrophasors

The power industry faces an ever-increasing demand for real-time measurements and for employing WAMSs that can be used to maintain or improve the stability of the system. Conventional SCADA/EMS stability control systems may not provide efficient solutions in the case of cascaded outages as these systems are designed to

26. A recloser is a protection device for electrical distribution networks. It combines a circuit breaker that trips if an overcurrent is detected (indicating a short circuit somewhere in a section of the network), with an electronically-controlled reclosing function that automatically restores power to the affected line if the fault clears itself quickly—which usually happens around 80 percent of the time.
27. American Gas Association Report No. 3, Orifice Metering of Natural Gas and Other Related Hydrocarbon Fluids. Founded in 1918, the American Gas Association (AGA) represents more than 200 local energy companies that deliver clean natural gas throughout the United States.
28. Any arcs or sparks or hot surface in electronic equipment has insufficient energy (heat) to start an explosion. Intrinsically safe electronic equipment is used in petro-chemical industry, coal mines, etc.

act locally—based on set operational parameters—and do not provide for dynamic coverage of a wide area power network [9].

WAMSs, on the other hand, provide for a dynamic coverage of the network and are able to handle cascaded outages through coordinated and optimized stabilizing actions. Techniques like phasor measurement units (PMUs)[29] are being used in WAMSs to complement conventional SCADA/EMS systems, thereby providing a wide area system view and increased system stability. In typical applications, PMUs are sampled from remote locations in the power system network and synchronized from the common time source of a GPS radio clock, which can provide a theoretical accuracy of synchronization better than 1 microsecond.

Synchrophasors provide a real-time measurement of electrical quantities from across the power system. Applications include wide-area control, system model validation, determining stability margins, maximizing stable system loading, islanding detection, system-wide disturbance recording, and visualization of dynamic system response. Synchrophasors are used to time-align data across systems and provide a true picture system-wide; they allow real-time power stability analysis and are critical in helping avoid cascading blackouts.

In addition, time-stamped synchronized measurements offer a tremendous benefit for protective relay applications. These real-time measurements represent actual system conditions at any given time and can be utilized in relay protection. The relays that are set to make a trip decision based on synchrophasor measurements constitute a class of adaptive or predictive protective devices [10].

The basic system building blocks are GPS satellite-synchronized clocks, PMUs, a phasor data concentrator (PDC), communications equipment, and application software. An important thing to remember is that the PMUs from multiple vendors can yield inaccurate readings, so it is recommended that all the PMU hardware, software, and protocols are made by the same manufacturer.[30]

A PDC forms a node in a system where phasor data from a number of PMUs or PDCs is correlated and fed out as a single stream to other applications. (See the simplified block diagram in Figure 1.11). The PDC correlates phasor data by time-tag to create a system wide measurement set.

The PDC provides additional functions as well: It performs various quality checks on the phasor data and inserts appropriate flags into the correlated data stream; it also checks disturbance flags and records files of data for analysis,

29. A phasor is a rotating vector represented by a complex number that contains both the magnitude and phase angle of the sine waves found in electricity. The phase or phase angle is the distance between the signal's sinusoidal peak and a specified reference and is expressed using an angular measure. The phasor magnitude is related to the amplitude of the sinusoidal signal. By definition they rotate counter-clockwise. Phasor measurements that occur at the same time are called synchronized phasors (synchrophasors), as are the PMU devices that allow their measurement. The concept of using phasors to describe power system operating quantities dates back to 1893 and Charles Proteus Steinmetz's paper on mathematical techniques for analyzing AC networks.

30. Synchrophasor standards have evolved since the introduction of the first one, IEEE Std. 1344, in 1995. IEEE Std. C37.118-2005 introduced measurement accuracy under steady state conditions as well as interference rejection. In 2009, the IEEE started a joint project with IEC to harmonize real-time communications in IEEE Std. C37.118-2005 with the IEC 61850 communication standard. These efforts led to the need to split the C37.118 into two different standards: IEEE Std. C37.118.1-2011 IEEE Standard for Synchrophasor Measurements for Power Systems and IEEE Std. C37.118.2-2011 Synchrophasor Data Transfer for Power Systems. New amendments were added in 2014.

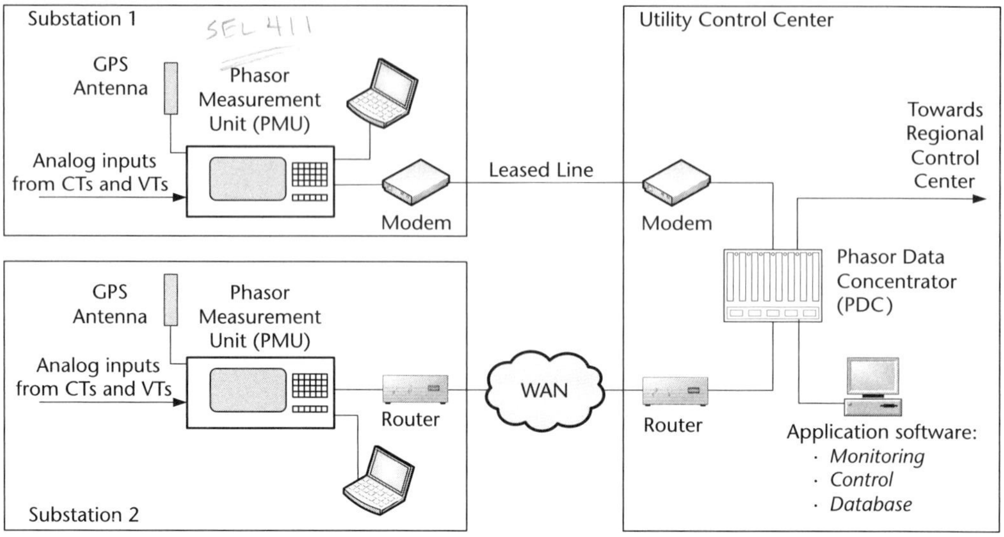

Figure 1.11 PDC and PMUs.

monitors the overall measurement system, and provides a display and record of performance. It can provide a number of specialized outputs, such as a direct interface to a SCADA or EMS system.

In order to ensure secure information transfer outside of the substation, the multilayered system architecture should include firewalls and encryption, and in addition, all unused ports on all devices should be disabled. The IEEE C37.118 data should be sent using UDP secure (UDP_S).[31]

1.3.4 Protective Relaying

Modern power system relays are electromechanical, electronic, and/or computer-based devices that protect power system equipment and apparatus from abnormal currents and voltages. The two fundamental relay functions are to isolate faulted sections of the power system while maintaining the power delivery capability in the rest of the power system. Figure 1.12 represents a simplified view of a protection relay.

The historical development of relays, which started in the early 1900s, can be classified into three major categories, electromechanical relays, static relays, and programmable relays, described as follows.

- Electromechanical relays: Early protective relays were designed to trip a circuit breaker when a fault was detected. They relied on coils operating on moving parts to provide detection of abnormal operating conditions such as overcurrent, overvoltage, reverse power flow, and over- and underfrequency. They are still being used by utilities, industrial, and commercial facilities.

31. UDP stands for User Datagram Protocol. By using UDP secure (UDP_S), the PDC sending the data neither expects nor accepts any incoming message streams, thus eliminating the possibility that a hacker could access the PDC.

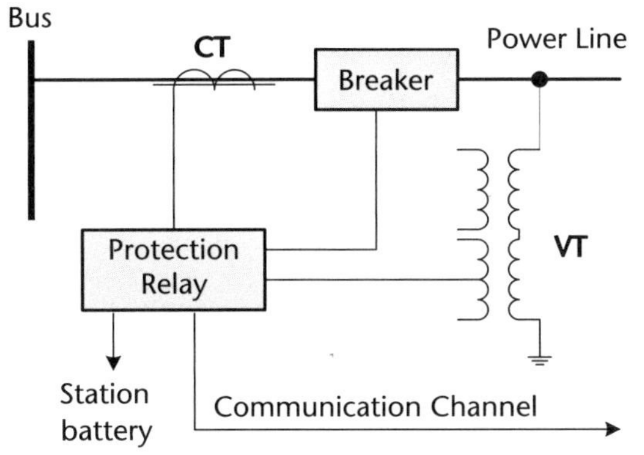

VT– Voltage Transformer
CT–Current Transformer

Figure 1.12 Single-line diagram of a protection relay.

They have limited capability and were packaged such that separate units, called elements, provided each control feature.

- Static relays: These use semiconductor devices like diodes, transistors, thyristors, integrated circuits and other electronic components. They are called static because there are no moving parts.
- Programmable relays: Today, microprocessor-based digital protection relays emulate the original devices and provide types of protection and supervision impractical with electromechanical relays. In many cases a single microprocessor relay provides functions that would take two or more electromechanical devices. With the presence of programmable microprocessors any relay characteristic that can be expressed mathematically is possible to realize in the field.

The need to act quickly to protect circuits and equipment as well as the general public often requires protective relays to respond and trip a breaker within a few milliseconds; therefore, it is necessary to properly maintain and regularly test protective relays.

Some automatic switching control provided by relays will reclose a breaker shortly after a trip in an attempt to quickly restore power to a circuit. Such reclosing operations are based on the experience that a high percentage of some types of faults are transitory and self-clearing after the line is de-energized. Relays have evolved to control, record, and report, communicate, and in some cases, adapt to, events in the power system.

First relay to use fiber optics for its channel was introduced in the late 1980s. It proved that relay system security and dependability is significantly improved by the use of fiber. Its only negative feature is that a pair of dedicated fibers is required to provide the communications. Protective relaying imposes the most stringent requirements on the communications channel, and therefore teleprotection channels are required to be fast, reliable, dependable, and secure.

Typically, there is a primary communications system for protection and data transmission, as well as a secondary (redundant) system with path diversity in case the primary system fails [11]. Although only fiber-optic (for example, OPGW),[32] and/or licensed microwave systems are considered for the primary communication channel, there is a number of other technologies that could be used for a secondary (redundant) communication system, such as PLC,[33] circuits leased from a telco (leased lines), and various hybrid solutions.

In some cases, two geographically separated routes are required to minimize the risk of a simultaneous interruption of both circuits, and many utilities will use a combination of private and public communications networks to develop the best solution for their needs. [12].

1.3.5 Voice Communications

Voice communications is a critical service for power systems operations especially during emergencies. Load dispatchers will require secure communications with power plant operators and substation operators during these situations to restore power supply with minimum disruption.

Use of public telecommunication voice services—fixed line or mobile—during supply interruption is not reliable, as network congestion and failure is high during power supply failures. Many power utilities prefer to have their own voice communications network and often use a private radio network but one that can be patched through to the public fixed and/or mobile network. Many utilities own and operate their own dedicated TDM or IP-based PABX systems for grid system operation. The network usually uses a three- or four-digit numbering system and supports voice communication between operators and maintenance crews, as well as within the company. Voice communication is crucial for all aspects related to the physical security of staff and plant.

A proper public address (PA) system covering the different areas of the power plant and substations is required not only for maintenance activities but also for the security of the installations. This system is critical in case of emergency situations where staff should be informed and/or accessed very quickly.

Project 25 (P25) was launched in 1988 as a step beyond the old-fashioned two-way voice contact between first responders and their dispatchers, with the dispatchers serving as the link to other agencies (including Red Cross and electric utilities) when necessary, generally over a telephone line. P25 is an open-architecture, user-driven group of standards that defines digital radio communications system architectures capable of serving the needs of public safety and government organizations. The P25 suite of standards involves digital land mobile radio (LMR) services for local, state/provincial and national (federal) public safety organizations and agencies.

Developed primarily for North American public safety services, P25 is a set of standards produced through the joint efforts of the Association of Public Safety Communications Officials International (APCO), the National Association of State

32. OPGW is a composite groundwire conductor that contains optical fibers in its central core.

33. PLC systems, operating on narrow channels between 30 and 500 kHz, are frequently used for high-voltage-line protective relaying applications. Messages are simple, one-bit messages using either amplitude- or frequency-shift keying, which tells the other end of a dedicated link to trip or to inhibit the tripping of a protective circuit breaker.

Telecommunications Directors (NASTD), selected federal agencies, and the National Communications System (NCS) and standardized under the Telecommunications Industry Association (TIA).

P25-compliant radios can communicate in analog mode with legacy radios and in either digital or analog mode with other P25 radios.

Terrestrial trunked radio (TETRA) is a digital trunked mobile radio standard developed by the European Telecommunications Standards Institute (ETSI). TETRA meets the needs of traditional professional mobile radio (PMR). TETRA protocol standards are a parallel effort, with much in common with P25, but the two are not compatible.

TETRA is a time division multiple access (TDMA) system with some additional features over P25; for example, in a TETRA system, if communication to the base station is lost, mobile radios can still communicate among themselves, a very useful feature in emergency situations. TETRA has been deployed in many countries around the world, using different frequency bands, and in the United States, it could be using the 450-MHz band in the near future.

1.3.6 Closed-Circuit Television

Closed-circuit television is widely used for surveillance and security. Video surveillance is one of the applications used to control people's access to premises. It is also used in some special process applications, such as the visual monitoring of water levels and animal intrusion. Visual confirmation of events or issues is particularly useful when personal safety is involved, such as verifying whether a breaker is open or closed before an operator enters an area.

Aside from preventing and investigating theft, unauthorized access, vandalism, sabotage, terrorism or other uncontrolled damage to critical infrastructure,[34] intelligent video monitoring systems sometimes utilize thermal cameras to automatically detect temperature increases in substation equipment, to detect anomalies, and to prevent serious failures before they happen.

The security applications support the automatic monitoring of defined areas or access, as well as the recording of scenes for future analysis. Fiber-optic and copper cabling, as well as radio connectivity can be used to back-haul traffic from locally and/or remotely installed cameras. The substation environment is harsh, and all the equipment requires careful design consideration of high levels of electromagnetic interference (EMI), temperature, vibration and shock, and even pollution.

1.3.7 Corporate Data Network

A corporate data network is used for enabling corporate IT systems that are necessary for business operations such as enterprise resource planning, customer information and billing systems, e-mail, management information systems, facilities management, and Internet access.

34. Critical infrastructure are systems and assets, whether physical or virtual, so vital that their incapacity or destruction may have a debilitating impact on the security, economy, public health or safety, environment, or any combination of these matters, across any Federal, State, regional, territorial, or local jurisdiction.

Business and administrative services do not require as stringent delay and availability performance as compared with the operational requirements of SCADA and protective relaying; however, their data network bandwidth requirements are usually much higher.

1.3.8 OPGW

Distances over which communication is required vary considerably—from only a few tens of feet, where the devices of interest are located in the same substation, to applications on high-voltage lines where the distance between devices may be hundreds of miles.

Figure 1.13 represents a typical high-level diagram of fiber-optic communication channel using OPGW. OPGW is colocated within an electric utility's own right-of-way, typically 60–150 feet above ground, protected by high-voltage electric facilities, and accessed by splice boxes. As depicted in Figure 1.13, it is common that the OPGW is transitioned to all-dielectric self-support (ADSS) fiber-optic cable at the entrance of the substation [13] to avoid situations that are hazardous to people and/or equipment.

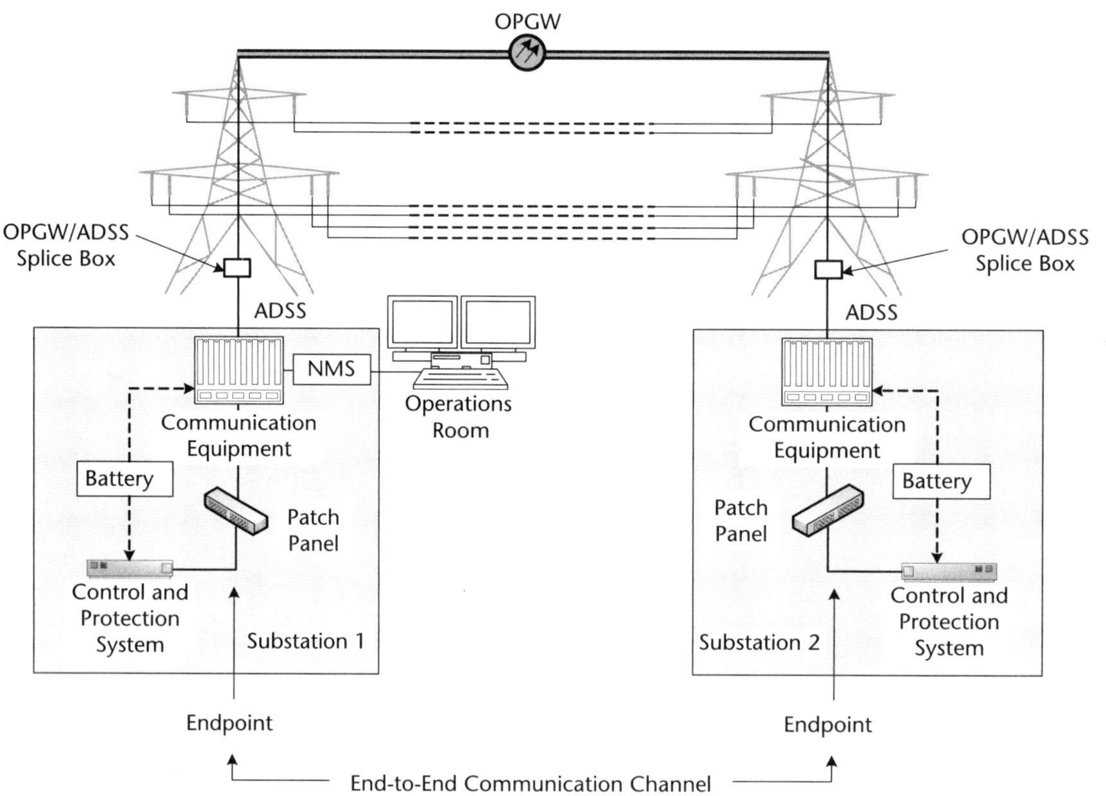

Figure 1.13 OPGW communication system.

1.3.9 Power Line Carrier (PLC)

1.3.9.1 About PLC

PLC systems have been used by electric power utilities since the 1920s for the transmission of vital information for the operation and protection of the electric power grid (i.e., voice, protection commands and control signals).[35] Thus, power lines are used not only to carry electrical power (at 50 or 60 Hz) but also to transmit communications signals, typically at much higher frequencies [14].

PLC is based on 1940s technology, and it is a point-to-point communications system using special devices directly coupled to the high-voltage transmission lines.[36] PLC involves sending information between a transmitter at one terminal station and a receiver at an adjacent station. In order to meet the varying requirements of power utilities when constructing a new power system or extending an existing one, PLC equipment must be compatible and of modular design.

The use of existing power lines for communications is a logical choice, because they can provide the most direct link for teleprotection (where high speed and low latency is crucial); they are reliable; and they are completely under the control of the power utility.

Power lines are a decent communications medium that can bridge very long distances (hundreds of miles) without a repeater, as long as we keep in mind potential challenges. There are three main types of disturbances on transmission lines that may affect communications: the well-known thermal noise; 60-Hz harmonics; and transient sources of noise, including lightning and nearby electromagnetic radiation sources. The transient sources of noise, especially lightning, are very unpredictable and can increase noise and introduce errors into the digital bit stream. [15].

1.3.9.2 Narrowband and Broadband PLC Systems

Generally, we can divide PLC systems into two groups: narrowband PLC, allowing communications services with relatively low data rates (up to 128 kbps) and ensuring realization of various automation and control applications as well as a few voice channels, and broadband PLC (BPL) systems, which allow data rates beyond 2 Mbps.[37]

The IEEE standard that defines low-frequency (below 500 kHz) narrowband power line communication and meets the requirements of the current and emerging smart grid and low bandwidth applications is IEEE 1901.2 [16].

Electric utilities are allowed to use PLC transmitters and receivers for control signals and information transmission in the 10–490-kHz band without obtaining a license from the FCC. However, PLC users are not protected from interference from licensed radio transmitters.

---

35. Although an idea of using power lines for transmitting voice messages appears to have occurred to a number of people and organizations in a number of countries, the first test and commercial operation of carrier telephony over power lines, as reported in the technical literature, took place in 1918 in Japan.

36. The PLC Forum is a leading international association that represents the interests of manufacturers, energy utilities and other organizations (e.g., universities, other PLC associations, and consultants.) active in the field of access and in-home PLC (power line communications) technologies

37. Data transmission at 2 Mbps or higher is considered “broadband.”

PLCs can also be used for home automation (e.g., remote lighting and appliance control and/or monitoring). Many such monitoring and control devices for home automation are available commercially today.

Since PLC does not require FCC licensing, the Utilities Telecom Council (UTC)[38] maintains a database that documents the use of the PLC frequencies in the United States. The frequencies are submitted by the power utilities with longitude and latitude information to help coordinate the use of the spectrum in the PLC range. This database is only available to members of the UTC.

BPL allows for the realization of a number of typical telecommunication services in parallel, such as telephony and Internet access. BPL represents a technology that enables electric utilities to provide support for high-speed data communications over their infrastructure into our homes and offices. Typically, BPL services operate by modulating in a carrier wave of between 1.6 and 80 MHz into electric power lines.

1.3.9.3 PLC Components Description

Figure 1.14 shows a typical end of a simple line terminal with the PLC equipment. Note that in Figure 1.14, the transmitted signal goes from the T/R transmitter via the coax cable to the line tuner, and then via high-voltage lead in wire from the tuner to the coupling capacitor device, which couples the signal into the transmission line. Conversely, the received signal reverses this path and arrives at the receiver of the T/R set [17].

A PLC coupling device, together with the associated coupling capacitor (CC) or capacitive voltage transformer (CVT), forms a filter that accepts the carrier

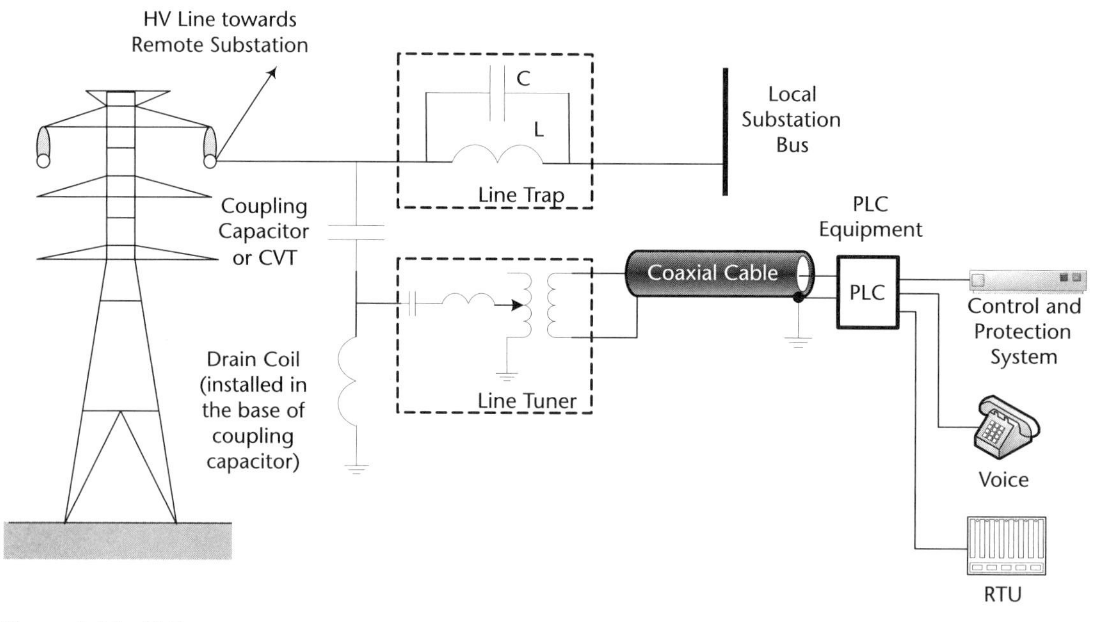

Figure 1.14 PLC.

38. http://www.utc.org/plc-forms-instructions (accessed July 29, 2015)

frequency signals and rejects the power system frequency. The following explanation of the principle of the coupling device assumes a single-phase coupling. (PLC signals can be coupled to one or more phases of the power system.)

A complete coupling comprises a line trap—to prevent the PLC signals from being short-circuited by the substation—and a coupling filter formed by the coupling capacitor and the coupling device (frequently, referred to as a coupling filter). Its purpose is to permit the PLC signal to pass but to reject the power system frequency and protect the communications equipment from the power system voltage as well as transient overvoltages caused by switching operations and atmospheric discharges.

Many factors will affect the reliability of a PLC channel; the goal is to get a signal level to the remote terminal that is above the sensitivity of the receiver, and with a SNR well above the minimum, so that the receiver can make a correct decision based on the information transmitted. If both of these requirements are met then the PLC channel will be reliable [18].

On short lines there are technical and financial reasons to use higher frequency and lower power. Care must be taken that the transmitter's spurious and harmonic output stays within the specification limits when the transmitter power output is set to some lower value. The equipment manufacturer should be consulted before making this adjustment.

In the United States, carrier frequency is 10 kHz–490 kHz (different than other countries). When planning a PLC system on frequencies of 250–500 kHz, a frequency coordination with local airport beacons must be carried out.[39] Airport beacons are licensed users of this frequency band, and PLC signals must not interfere with these users.

1.3.10 Broadband Power Line Carrier (BPL)

1.3.10.1 BPL and Smart Grid

Historically, power utilities have used alternating current (AC) power line distribution facilities to carry information by coupling RF energy to AC electrical wiring in houses or buildings. In the past, these devices have operated on frequencies below 2 MHz with limited communications capabilities. Due to power line characteristics it has been difficult to achieve dependable high-speed communications. However, technological advancements have resulted in the development of new systems that have overcome these technical obstacles. Trials have demonstrated that high-speed communication voice and data services can be achieved using the existing MV and low-voltage (LV) power distribution grid.[40]

Electric utilities are not entering the communications business; they leave that part of BPLC to a partner, perhaps an Internet service provider (ISP), a competitive local exchange carrier (CLEC), or a long-distance company looking for an alternative last mile path to their customers. In addition to broadband Internet connec-

39. For example, nondirectional navigational radio beacons (NDBs) for maritime and aircraft navigation occupy a band from 190 to 435 kHz.

40. Despite the spread of broadband technology, many areas of the world that don't have access to high-speed Internet, and broadband could potentially be provided through power lines.

tions, these broadband distribution systems are also intended to provide voice, video, surveillance systems, and entertainment (gaming).

The current focus of most electric utilities is using BPL for an intelligent electric distribution grid (smart grid) and utilities metering (electricity/water/gas) services. This could result in lower electric power costs, less pollution, and greater reliability and security—essentially, a more intelligent electric power grid.

The new systems use spread spectrum or multiple carrier techniques with built-in algorithms to circumvent the noise in the power line. For example, orthogonal frequency division multiplexing (OFDM)[41] is used to distribute the BPL signal over a wide bandwidth using many narrowband subcarriers. OFDM is much more spectrally efficient than FDM because it reduces the required bandwidth by squeezing subcarriers tightly together until they actually overlap with each other. This is accomplished by keeping the subcarriers orthogonal in the complex domain so that they do not interfere with each other.

1.3.10.2 BPL Principle of Operation

The power distribution grid is made up of a number of components aimed at delivering electricity to customers, and includes overhead and underground MV and LV power lines and associated transformers. MV power lines carry electricity in the range of 12.5–36 kV, and the voltage is decreased by step-down transformers to provide LV power to houses and buildings. In general, the LV power lines carry electricity at 120/240V or 347/600V.

Figure 1.15 shows IP data from the Internet being injected at the HV/MV substation, traveling through the electrical grid, and being accessed by consumers using a BPL modem at their premises. The BPL equipment injects the IP data as an RF signal into the overhead or underground power lines using ferrite cores. The RF

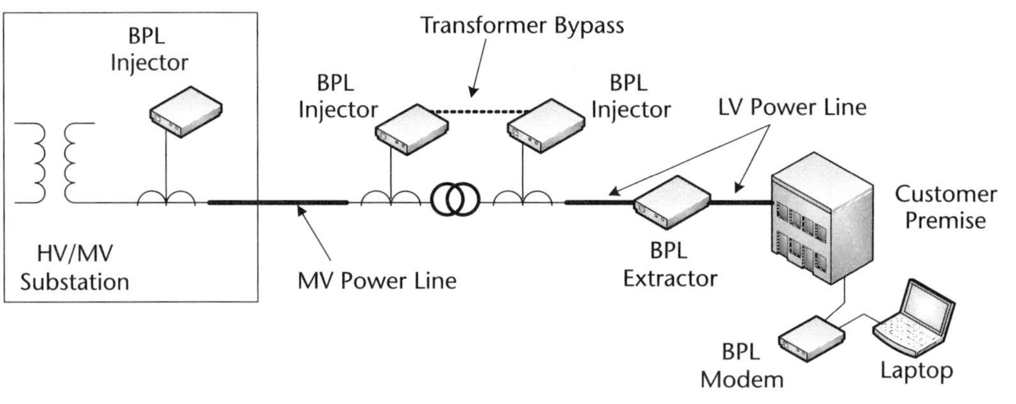

Figure 1.15 BPL architecture.

41. OFDM has been adopted for numerous technologies including: asymmetric digital subscriber line (ADSL), IEEE 802.11a/g, IEEE 802.16a, digital audio broadcast (DAB), Digital Terrestrial Television Broadcast, 4G wireless networks, IEEE 802.11n, IEEE 802.16, and IEEE 802.20. There is no single OFDM standard that has been universally adopted; there are similar but different versions of OFDM and several proprietary implementations.

signal is typically injected at the MV substation, bypasses LV transformers, and is delivered to the end users' electricity sockets in their home or business. If necessary, repeaters are used to reamplify and repackage the signal because the signal gets attenuated as it travels along the power line.

BPL systems are comprised of different components that function together to deliver broadband services to customers. There are two basic types of BPL solutions, access and in-building. Typically, access BPL connects the MV substation to the consumer's premises, whereas in-building BPL is a LAN replacement.

BPL systems are designed to send information within parts of the 2–80-MHz frequency range along unshielded power lines, which results in the unintended emission of RF energy and may impact the current radio-communication environment by creating interference to existing services. There are a number of authorized radio services in the 2–80-MHz frequency range: amateur radio, fixed mobile, maritime mobile, aeronautical mobile, fixed broadcasting, space research, radio astronomy, and aeronautical radio navigation. Their users include public safety organizations, federal government agencies, aeronautical navigation licensees, amateur radio operators, international broadcasting stations, and citizens band (CB) radio.

There are several BPL technologies and more BPL industry alliance-backed or proprietary solutions. Not only do these technologies not interoperate, but simultaneous operation of these technologies using the same frequency bands in proximity may cause harmful mutual interference, which may lead to severe performance degradation or even malfunctions in both smart grid and/or home networking applications. Harmonization of IEEE 1901 and ITU-T G.9903 and G.9904 standards is in progress, while harmonization with ISO/IEC 12139-1 is an open issue. [19]

1.3.11 Leased Lines

Sometimes it will be necessary to lease communications circuits from the common carrier (telco). Figure 1.16 illustrates one of the possible scenarios of leased transport facilities and what the end-to-end circuit might look like.[42]

Local exchange carriers (LECs) operate within specific franchised service areas, basically service monopolies, called local access and transport areas (LATAs). In general, each LATA contains a single major metropolitan or market area that may not necessarily match up with existing state, province, or even area code borders. Most incumbent LECs (ILECs) operate across a number of LATAs.

The border of a LATA defines where local service ends and long-distance service begins. A LATA is associated with one or more telephone area codes. LECs may be one of the incumbent carriers and a competitive access provider. The local carrier (i.e., local loop) typically has several switching offices [central offices (COs)] within the same LATA.

If the leased digital T1 circuit crosses multiple states, it will require the cooperation of multiple telecommunications companies. Telephone calls and telecommunications circuits between two parties in different LATAs (InterLATA) must be,

42. Dedicated four-wire circuits (based on TDM) between two sites, available for decades from local telecom companies, based on some predictions are slowly going to be phased-out by 2020. Telcos will offer leased circuits using "cloud" and virtual private networks (VPNs) that in most cases are sufficient for voice and low-priority data but do not necessarily guarantee performance required by teleprotection.

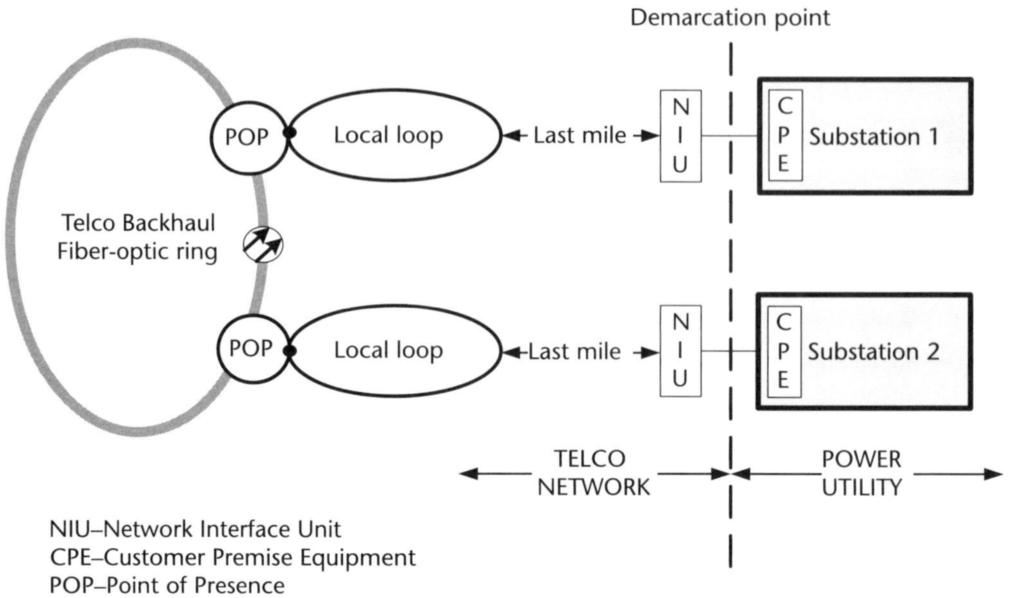

Figure 1.16 Leased lines.

by law, transported by a long-distance carrier. An interexchange carrier (IXC) is any long-distance provider that provides services between the LECs. The LECs are required to provide a point-of-presence (POP) (i.e., an interface for the IXCs).

1.4 Power Grid and the Need for Modernization

1.4.1 About Power Grid

Power grid is a network for the transmission of electricity throughout a state or region. The term "grid" usually refers to the transmission lines; however, the power system is designed as an integrated system that specifically relies on generation and transmission to move power from location to location.

The transmission grid is designed as a network, meaning that the connections allow two-way power flows. In contrast, local distribution systems are generally designed for power to flow one way, from the transmission lines to end users. The term "radial" is applied to these one-way transmission and distribution elements, with an exception being transmission lines that link generation to the bulk power network, and obviously power flows one way along these lines. They are called integrating transmission lines because they integrate generators into the grid.

The United States' power transmission network is divided into three major systems (western, eastern, and Texas), which are electrically isolated from each other. Within each of these networks (called "interconnections"), transmission systems are operated at a regional or utility-area level by "control centers." There are roughly 140 control centers in North America, and each control center manages power flows within its own boundaries and coordinates flows across boundaries with adjacent control areas. If there is a major system failure, the network breaks down into component systems based on these control areas. As a result, the entire country (or interconnection) is saved from a blackout.

1.4.2 Modernization of the Power Grid

The Energy Independence and Security Act of 2007 established a goal to modernize the nation's electricity system and assigned to NIST[43] the primary responsibility to coordinate development of a framework to achieve interoperability of smart grid devices and systems. Power grid modernization should ensure backward compatibility with existing technology as much as possible [20].

NIST's concept model provides a high-level, overarching perspective of major relationships across different domains of power-grid systems like generation, transmission, distribution, and energy sources, as well as users with the capability to make decisions and exchange information with other users. This concept model defines information flow between different domains and users within the smart grid.

Severe storms such as Hurricane Sandy in October 2012 produce long-term outages resulting in billions of dollars in infrastructure damages and lost business revenue.[44] Therefore, it is important to consider extreme events and their mitigation: natural and man-made events such as EMI, geomagnetic storms, high-altitude nuclear detonations, and severe weather. They all pose a potential threat to grid reliability and resiliency.

EMI can degrade or disable effective two-way communication and control on the grid; geomagnetic storms can damage large transformers and other equipment through induced currents, and a high-altitude nuclear detonation would produce an electromagnetic pulse capable of damaging unprotected integrated circuits as well as large electrical equipment.

In the past, electric utilities almost exclusively communicated with the public through mass media, such as TV and radio. Today, different communications channels are available for utilities to share restoration notifications with a customer's personal device (e.g., cell phones, Twitter, Facebook, and e-mail)[45] while advanced metering infrastructure (AMI) has enabled utilities to "ping" meters and inform customers when their power has been restored. When it comes to dealing with the public, electric utilities are creating internal and external communications messaging protocols and processes, utilizing grid modernization technologies to reduce conflicting messages during a major power outage event.

Aside from the United States, many other countries are trying to modernize the electric grid on different continents, and they have launched significant efforts to encourage the development of the smart grid in their own countries and regions. As countries move forward with individual initiatives, internationally coordinated and harmonized smart grid efforts will be essential.

---

43. Founded in 1901, NIST is a nonregulatory federal agency within the U.S. Department of Commerce.

44. GridWise Alliance, Improving Electric Grid Reliability and Resilience: Lessons Learned from Superstorm Sandy and Other Extreme Events, June 2013.

45. During Hurricane Katrina restoration, texting proved much more effective than voice calls due to lower bandwidth requirements and the "store and forward" capability, which enabled text messages to get through busy cellular networks.

1.4.3 Smart Grid and Utility Communications Architecture

In January 1998, the United States deregulated the power utility industry.[46] As a result, customer's energy usage must be closely watched so that utilities can learn about their needs and desires and be able to offer them additional services and rate structures. The implementation of a network to monitor the "real-time loads" will greatly increase the profitability of the utility company.

The power utility network has become much more sophisticated in recent years. This can be attributed to the growth in distributed renewable generation with its intermittency issues, the reverse billing systems associated with net metering, and the volume of data that smart meters are sending back to utilities from customers' homes and businesses.

There are many different definitions of smart grid, including the following:

> Smart grid is an automated, widely distributed energy delivery network characterized by a two-way flow of electricity and information, capable of monitoring and responding to changes in everything from power plants to customer preferences to individual appliances.

Regardless of the formal definition, the purpose of SG is to identify and resolve faults on the electrical grid, to automatically self-heal the grid, to monitor power quality and manage voltage, to identify devices or subsystems that require maintenance, to help consumers optimize their individual electricity consumption (minimize their bills), and to enable the use of smart appliances that can be programmed to run on off-peak power. All this will, of course, help to physically maintain the grid and troubleshoot the problems in more efficient way than before (smart grid = electric grid + intelligence). Figure 1.17 shows the high-level diagram with the main building blocks of the modern SG architecture.

Using automation to improve asset management, grid maintenance, and workforce dispatch is a key capability through which the smart grid improves operational efficiency; a modern and robust communications network is central to supporting those operations. At the same time, the smart grid, the communications network, and the information the network carries must be protected from unauthorized access at all times—for public safety, personal privacy, and regulatory compliance.

Some people assume the smart grid[47] is a revolutionary change to the operation of the electric grid, without realizing that many of the concepts and even technologies existed long before the term smart grid was even used. In reality, it is just an incremental step in the long evolution of adding automation to the electric grid. While the term smart grid is generally associated with electric utilities, the same concepts can be applied by gas, water, and wastewater utilities [21].

---

46. Deregulation means that the customer has the power to choose the company to buy electricity or natural gas from. No matter which company a customer selects, the electricity and natural gas is then transmitted home by the utility that owns the entire infrastructure and makes repairs when things go wrong.

47. National Academy of Engineering (NAE) named 20 engineering achievements that had the greatest impact on quality of life in the twentieth century. Electrification was ranked first, electronics was fifth, telephony ranked ninth, and the Internet was thirteenth. When the NAE ranks the twenty-first century's greatest achievements, it is likely that these achievements will still be on the top of the list but this time as a single category: the Smart Grid.

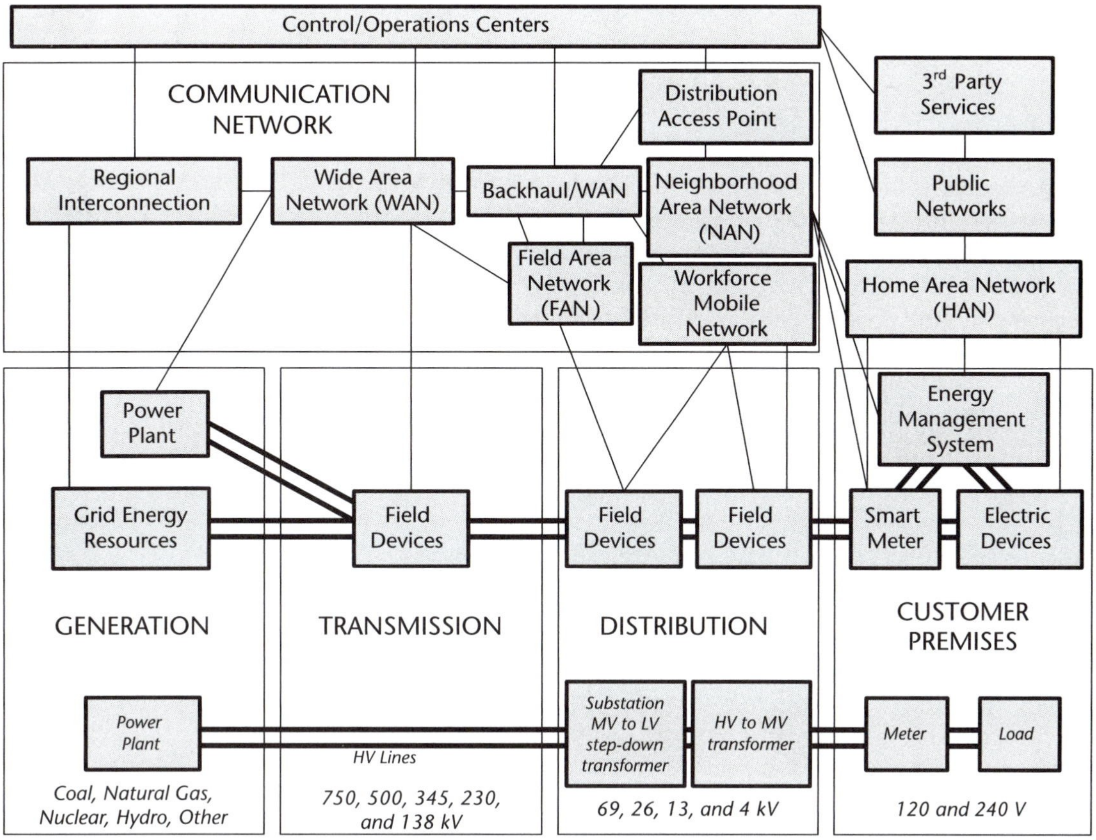

Figure 1.17 Power grid and communications network architecture.

1.4.4 Wireless Sensor Networks (WSNs)

Different from the existing power grids, in the smart grid, online and reliable information is the main factor for the reliable delivery of energy from the generation units to the users. The utility industry is modernizing its meter reading and data collection system with AMI.

Generally, the smart grid network consists of the following three segments: home area networks (HANs), which connect smart meters with on-premise appliances, distributed renewable sources, and electrical vehicles; neighborhood area networks (NANs), which carry information between customer premises and aggregation points; and WANs, which serve as the backbone for communication between the utility data center and aggregation points.

Recent advances in embedded systems and WSNs made it possible to implement low-cost monitoring and diagnostic systems for smart grids. These systems receive information from wireless sensor nodes, which monitor critical smart grid equipment and respond to the changing conditions in a proactive manner. Hence, WSNs have been recognized as a promising and complementary technology for various smart grid applications.

At a substation level, mesh radio networks have significant potential for transformer health monitoring, line temperature monitoring, and fault current indication. WSNs offer an advantage in their strong ability to adapt to harsh environments, their large area coverage, and their self-organization and self-configuration. WSNs are not single-technology but represent complex systems using various technologies from physical communication layers to application programs.

The core of these efforts is the IEEE 802.15.4 standard, which targets low-data rate applications requiring wireless interconnections between measurement, analysis, and control devices. The data packets can travel from the source to destination in one or more hops, so the routing protocol in WSN plays an important role in ensuring that data packets can travel over multiple hops from source to destination nodes. There are four major categories of routing protocols that have been proposed so far, including flooding, clustering, geographical routing, and self-organizing protocols, and they all have advantages and disadvantages, depending on the application [22].

Chapter 4 provides more information on radio networks and technologies that can be utilized to create a WSN.

1.4.5 Communications Technology Adoption

As mentioned earlier, utilities also have numerous communications platforms and networking protocols operating in their networks today. The majority of network communications still rely on private networks. The most widely deployed communications by utilities are leased telephone lines and leased T1/E1 circuits, fiber-optic systems, copper wires, cellular, and microwave networks. Legacy systems like PLC are also used.

Although these results reflect the state of communications in the global utility industry, it's important to note that there are regional differences in technology adoption. For example, WiMAX has been widely deployed in South America, while PLC is a dominant technology for access communications in Europe. Neither technology has a significant presence in the rapidly evolving smart grid markets in North America or Asia.

Utilities have several main concerns related to operating their current networks and making additional investments in communications. The foremost problem is the challenge to integrate and operate vendor-specific networks, while standards and interoperability also pose a significant challenge for utilities. Integrating these networks is a challenge that will persist with future versions of software and hardware. Network security is probably the third most important topic when it comes to utility communication networks.

1.5 Review Questions and Problems

1. Define transmission network in telecommunications.
2. List and compare different transmission media from technical and cost perspectives.
3. Loss of the radio signal, L, propagating through air, is determined by the fundamental Friis formula and depends on speed of light *c*, distance *d*, and frequency *f*:

$$L = \left(\frac{4\pi \cdot d \cdot f}{c} \right)^2$$

Show this formula for the radio signal loss (in decibels) using logarithms. Show the result as a sum of three logarithmic terms (constant term + frequency-dependent term + distance-dependent term).

$$L_{dB} = 10 \log L$$

(The answer can be found in Chapter 4.)

4. List and discuss some of the applications for the free-air laser communications. What are the main potential challenges of this technology?
5. What are the main contributors to the system latency for the fiber-optic link and microwave link?
6. Let's assume that PLC equipment uses a carrier frequency of 100 kHz. With the value of the inductance in the line trap (parallel LC circuit) equal to 0.25 mH, calculate the value of the required capacitance.

$$Z = \frac{j\omega L \times \dfrac{1}{j\omega C}}{j\omega L + \dfrac{1}{j\omega C}} = \frac{\dfrac{L}{C}}{j\left(\omega L - \dfrac{1}{\omega C} \right)}$$

Z will approach infinity and block 100 kHz signal when:

$$\omega L = \frac{1}{\omega C} \Rightarrow C = \frac{1}{(2\pi f)^2 L} = \underline{10\ nF}$$

   A. Calculate the resonance frequency of the parallel LC circuit with $C =$ 12 nF and $L = 0.22$ mH. (Solution: $f = 98$ kHz.)

7. How many watts of power represent 175 mW, 16 dBm, and 2.5 dBW? (Solution: 0.175W, 0.03981W, 1.7783W.)
8. A communications system that is transmitting signal of 5 dBm experienced losses of 17 dB and gain of 10 dB. What is the final power at the receiver? (Solution: –2 dBm)
9. Add a power of 25 dBm to a power of 35 dBm. (A word of caution about decibels per meter: You cannot add decibels over one milliwatt directly to decibels over one milliwatt!) The two powers must first be converted to milliwatts, then added, and the sum reconverted to decibels over one milliwatt. (Solution: 35.4 dBm.)
10. Synchronized phasors (synchrophasors, for short) provide a real-time measurement of electrical quantities in the power system and are used for wide-area control, system model validation, determining stability margins,

maximizing stable system loading, detection, system-wide disturbance recording, and visualization of dynamic system response.

A. Research and describe the functionality of the main building blocks of the system, including PMUs, a PDC, communications equipment, and visualization software.

B. Describe the importance of GPS satellite-synchronized clocks for the operation of the system. Special attention should be paid to loss of time signal and time quality problems in case of malicious hijacking/interference of the GPS signal.

References

[1] Dutton, Harry J. R., *Understanding Optical Communications*, IBM, 1998.

[2] www.greatachievements.org (accessed April 29, 2015).

[3] Mackay, S., et al, *Practical Telecommunications and Wireless Communications: For Business and Industry,* Burlington, MA: Practical Professional Books from Elsevier, 2004.

[4] Goleniewski, L., *Telecommunications Essentials,* Second Edition, Addison-Wesley Professional, Boston, MA: Pearson Education, 2006.

[5] Freeman, L. R., *Telecommunication System Engineering*, 4th Edition, New York: Wiley-Interscience, 2004.

[6] Federal Information Processing Standards, *Publication 197, Announcing the Advanced Encryption Standard (AES),* November 26, 2001.

[7] Barker W. C., and E. Barker, "Recommendation for the Triple Data Encryption Algorithm (TDEA) Block Cipher," Computer Security Division, Information Technology Laboratory, National Institute of Standards and Technology, revised January 2012.

[8] Ekanayake, J., et al, *Smart Grid Technology and Applications,* Chichester, West Sussex, UK: John Wiley & Sons, Ltd., 2012.

[9] Naduvathuparambil, B., et al., "Communication Delays in Wide Area Measurement Systems," IEEE, 2002.

[10] *Use of Synchrophasor Measurements in Protective Relaying Applications*, Power System Relaying Committee Report of Working Group C-14 of the System Protection Subcommittee, April 8, 2013.

[11] NERC, *"Reliability Standards for the Bulk Electric Systems of North America,"* 2008.

[12] NERC/WECC, *"Planning and Operating Criteria,"* 2007.

[13] Siemens, "Power Engineering Guide," Edition 7.0, 2012.

[14] Anatory, J., and N. Theethayi, *Broadband Power-line Communication Systems—Theory and Applications,* Southampton, UK: WIT Press, 2010.

[15] Bakhoum, E., "S-Parameters Model for Data Communications Over 3-Phase Transmission Lines," *IEEE Transactions on Smart Grid,* Vol. 2, No. 4, December 2011.

[16] IEEE P1901.2/D0.08.00, *Draft Standard for Low Frequency (Less than 500 kHz) Narrow Band Power Line Communications for Smart Grid Applications,* May 2013.

[17] IEEE 57th Annual Conference for Protective Relay Engineers, *Special Considerations in Applying Power Line Carrier for Protective Relaying,* IEEE Power Systems Relaying Committee Special paper, Relaying Communications Subcommittee, Working Group H9, College Station, Texas, 2004.

[18] IEEE Std. 643-2004, Guide for Power-Line Carrier Applications.

[19] NISTIR 7862, Guideline for the Implementation of Coexistence for Broadband Power Line Communication Standards, June 2012.

[20] NIST, *Framework and Roadmap for Smart Grid Interoperability Standards,* Release 3.0, 2014.

[21] Bariso, R., "Strategies for Building and Operating a Secure, Scalable and Reliable Smart Grid," *Electric Energy T&D Magazine,* May/June 2012.

[22] Majumder, R., et al., "Power Sharing and Control in Distributed Generation with Wireless Sensor Networks," *IEEE Transactions on Smart Grid,* Vol. 3, No. 2, June 2012.

CHAPTER 2

Communications Fundamentals

2.1 Analog and Digital Signals

An analog[1] signal is any continuous signal for which the time-varying feature (variable) of the signal is a representation of some other time varying quantity (i.e., analogous to another time-varying signal). Analog is usually thought of in an electrical context; however, mechanical, pneumatic, hydraulic, and other systems may also convey analog signals. Electrically, the property most commonly used is voltage followed closely by frequency, current, and charge.

Any information may be conveyed by an analog signal; often such a signal is a measured response to changes in physical phenomena, such as sound, light, temperature, position, or pressure, and is achieved using a transducer. For example, in sound recording, fluctuations in air pressure (that is to say, sound) strike the diaphragm of a microphone, which induces corresponding fluctuations in the current produced by a coil in an electromagnetic microphone, or the voltage produced by a condenser microphone. The voltage or the current is said to be an "analog" of the sound.

Analog signals are continuously variable and can be transmitted over various media (copper wire, fiber optic, space). In nature, signals and physical changes are always analog with a reasonably small rate of change.

Examples of some analog signals and their bandwidths are listed as follows:

- Speech bandwidth: 100–7,000 Hz;
- Telephone bandwidth: 300–3,400 Hz;
- Video bandwidth: 4 MHz.

Digital signals change instantly from one value to another (usually 0V and 5V). A digital signal is a sequence of logical 1's and 0's.

Figure 2.1 depicts what is known as plain old telephone service (POTS), an analog telephone service implemented over copper twisted pair wires and based on the Bell telephone system. This system connects homes and businesses to neighborhood central offices. A central office (CO) is ultimately connected to other offices

1. Derived from the Greek word *analogos* (i.e., proportional to the original).

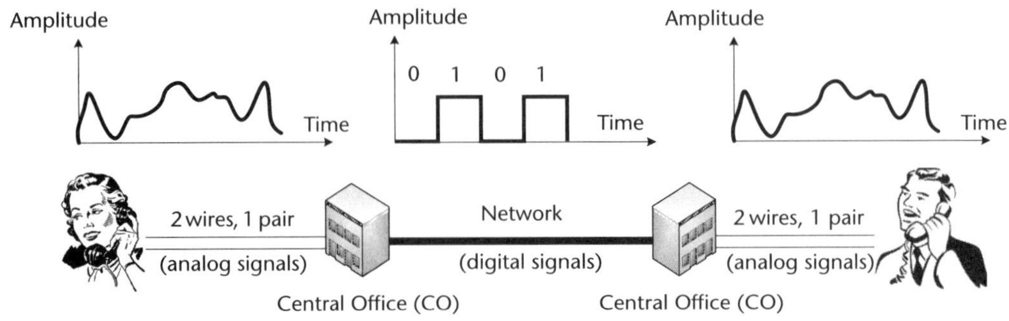

Figure 2.1 Analog and digital signals.

and a long distance facility. Due to its reliability, POTS is used more widely than any other telephony system.

Analog signals or voice are transmitted to the nearest CO and converted into digital signals. Digitized signals travel through the network to far-end CO, where they are converted back to analog signals and sent over a short distance to homes and offices.

A high-pitched voice contains mostly high frequencies, while a low-pitched voice contains low frequencies. A loud voice contains a high-amplitude signal, while a soft voice contains a low-amplitude signal.

Analog signals can be combined (i.e., multiplexed) by combining them with a carrier frequency. When there is more than one channel, the process is called frequency-division multiplexing (FDM). FDM was used extensively in the past but now has generally been replaced with the digital equivalent, called time-division multiplexing (TDM).

Digital signals used in communications systems today are binary signals, which can only take on one of two values: 0 or 1, commonly known as bits.[2]

Both analog and digital systems are subject to limitations in resolution and bandwidth. As analog systems become more complex, effects such as nonlinearity and noise ultimately degrade analog resolution to such an extent that the performance of digital systems may surpass it. In analog systems, it is difficult to detect when such degradation occurs. Similarly, as digital systems become more complex, errors can occur in the digital data stream. However, in digital systems, degradation can be detected and easily corrected.

To understand bits, we first look at digits. A digit is a single place that can hold numerical values in the range of 0 to 9. Digits are normally combined together to make larger numbers. For example, 957 has three digits; 7 is holding the 1's place; 5 is holding the 10's; and 9 is holding the 100's place. Each digit is placeholder for the next higher power of ten. Each place can have 10 different values; this implies that digits use a base-10 number system.

Bits use a base-2 (binary) number system and computers and communication systems use binary number system because it is easy to represent 1's and 0's using semiconductor switches. (Electric current either flows or it doesn't.) A binary digit is a sequence of 0's and 1's that represent numbers, for example:

2. Bit stands for Binary digIT.

```
0=0        2=10       3=11        4=100      5=101
6=110      7=111      8=1000      9=1001     10=1010
13=1x2^3+1x2^2+ 0x2^1+ 1x2^0= 1101
```

In a digital communications channel, the information can be represented by individual data bits that are encapsulated into multibit message units. Bits are typically grouped together into eight-bit combinations, called bytes (or octets). With eight bits in a byte, a byte can represent values ranging from 0 (00000000) to 255 (11111111).

Multiple bytes can be grouped to form a frame or other higher-level message unit; these types of multiple levels of encapsulation facilitate the handling of messages in a complex digital communication network.

2.2 Bandwidth

The quantity of information a channel can handle in a given period is determined by its ability to handle the rate of change of the signal (i.e., its frequency). The frequency of an analog signal varies between a minimum and a maximum value and the difference between those two frequencies ($f_H - f_L$) is the bandwidth of that signal. The bandwidth[3] of an analog channel is the difference between the highest and lowest frequencies that can be reliably transmitted over the channel (Figure 2.2).

Bandwidth is normally specified in terms of those frequencies at which the signal has fallen to half the power (or 0.707 of the voltage) relative to the midband frequencies, referred to as –3 dB points. In this case the bandwidth is known as the –3-dB bandwidth.

Digital signals are made up of a large number of frequency components[4], but only those within the bandwidth of the channel will be able to be received. Therefore, the larger the bandwidth of the channel, the higher the data transfer rate can be achieved and higher frequency components of the digital signal can be trans-

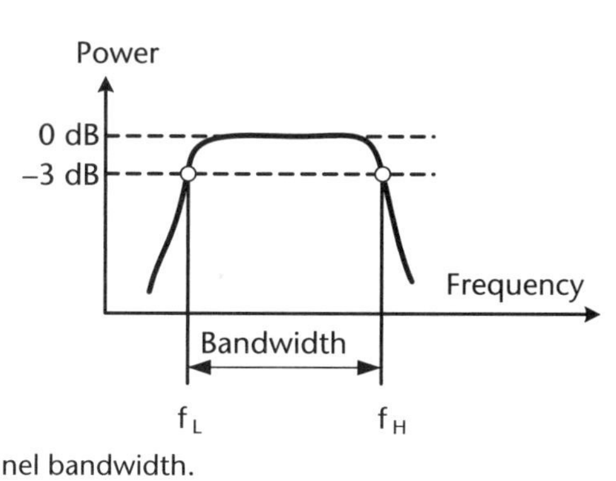

Figure 2.2 Analog channel bandwidth.

3. Analog bandwidth is sometimes called frequency bandwidth, spectral bandwidth, RF bandwidth, or signal bandwidth.

4. The Fourier transform is a mathematical tool that breaks a waveform (a function or signal) into an alternate representation, characterized by sine and cosine. The Fourier transform shows that any waveform can be rewritten as the sum of sinusoidal functions with different frequencies.

ported, resulting in a more accurate reproduction of the transmitted signal at the receiver.

2.3 Analog-to-Digital Conversion (ADC)

In the real world, most data is characterized by analog signals, while modern electronic equipment and computers deal with digital signals and binary 1's and 0's. Quite often we have to convert signals from analog to digital and back. See Figure 2.3.

Conversion of analog to digital involves three steps: sampling, quantization, and encoding. The sampling process converts an analog waveform into a sequence of discrete samples that are usually spaced uniformly in time. Quantization is the approximation of the instantaneous analog value. The quantization is done between the maximum amplitude value and the minimum amplitude value. In encoding, each approximated value is then converted into a binary format of 1's and 0's the computer can then recognize.

There are two parameters to control in converting the analog signals to the digital signals:

- Sampling rate, f_s: Controls the number of samples taken in a second, and it is based on sampling theorem described below;
- Sampling precision, N: Controls the number of different gradations (quantization levels) for the sampling process.

The quantization process is the limiting of a finite level measurement of a signal sample to a specific preset integer value. This means that the actual level (amplitude) of the sample may be greater or smaller than the preset reference level with which it is being compared. The difference between these two levels, called the quantization error, is compounded in the decoded signal as quantization noise.

2.3.1 Nyquist Theorem

The Nyquist theorem, also known as the sampling theorem, is a basic principle in the digitalization of analog signals. For ADC to result in a faithful reproduction of the signal, slices, called samples, of the analog waveform must be taken frequently. The number of samples per second is called the sampling rate or sampling frequency.

Any analog signal consists of components at various frequencies and this can be shown using Fourier analysis. The simplest case is the sine wave, in which all the signal energy is concentrated at one frequency. In practice, analog signals usually

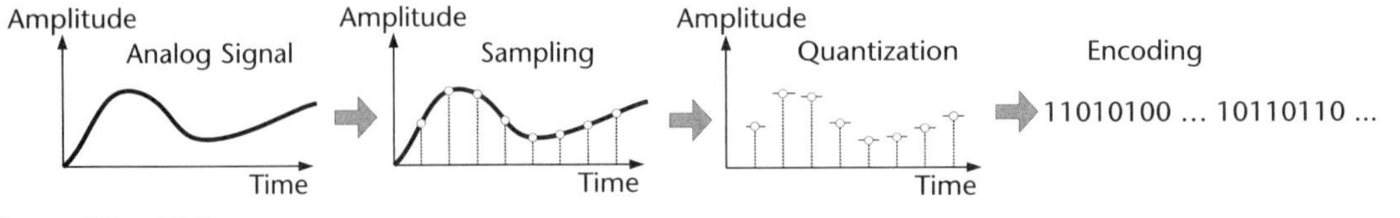

Figure 2.3 ADC.

have complex waveforms, with components at many frequencies. The highest frequency component in an analog signal determines the bandwidth of that signal.

Aliasing (a false lower frequency component) arises when a signal is discretely sampled at a rate that is insufficient to capture the changes in the signal. The Nyquist sampling theorem[5] simply states that the sampling frequency should be at least twice the highest frequency contained in the signal in order to avoid aliasing. In mathematical terms:

$$f_s \geq 2f_c \tag{2.1}$$

where f_s is the sampling frequency (how often samples are taken per unit of time or space), and f_c is the highest frequency contained in the signal.[6]

In the United States, the TDM system is known as the Tier 1 (T1) system, in which an analog voice channel is sampled 8,000 times per second (twice the maximum speech frequency of about 4,000 Hz) and each sample is encoded into a seven-bit byte. Twenty-four such channels are multiplexed together on two copper pairs and transmitted at a bit rate of 1.544 Mbps. T1 in North America (E1 = 2.048 Mbps in the rest of the world) remains an important method of transmitting voice and data in the public switched telephone network (PSTN) and elsewhere [1].

2.3.2 Signal-to-Noise Ratio (SNR)

Any signal transmission consists of useful signal and noise components. Noise is an unwanted disturbance on the communication channel and can be caused by many different mechanisms.

There are internal and external noise sources (e.g., electronic components, wiring and electric cables, motors, RF signals, and spark plugs). The strength of the transmitted signal is normally much larger than the noise signal. However, as the signal travels through the channel and is attenuated, its level can approach that of the noise. When the level of the wanted signal is not significantly higher than the background noise, the receiver cannot separate the data from the noise and communications errors occur.

An important parameter of the communications channel is the ratio of the power of the received signal (S) to the power of the noise signal (N), or the SNR, which is expressed in decibels:

$$S/N_{dB} = 10\log\frac{S}{N} \tag{2.2}$$

where

5. The quantity $2f_c$ is called the Nyquist rate, and transmitting at the limiting pulse rate of $2f_c$ pulses per second as signaling at the Nyquist rate. The reciprocal of Nyquist rate is called Nyquist interval. Nyquist published his results in 1928 as part of his paper "Certain topics in Telegraph Transmission Theory".

6. Explanation of 44.1 kHz CD sampling rate - the CD sampling rate has to be larger than about 40 kHz to fulfill the Nyquist criterion that requires sampling at twice the maximum analog frequency, which is about 20 kHz for audio.

S = signal power in watts or milliwatts;

N = noise power in watts or milliwatts.

The higher the signal level and the lower the noise level, the better the SNR and, therefore, the better the communication channel performance. It is also important to remember that the SNR sets an upper bound on the achievable data rate in digital communications systems.

2.3.3 Shannon Theorem

So far we have discussed ideal communications channels; in reality, a communications channel is never a perfect medium—there are always different impairments affecting it. At the moment, we will call all these different impairments noise; a simplified representation is shown in Figure 2.4.

In information theory, the noisy-channel coding theorem (stated by Claude Shannon in 1948) [2], establishes that for any given degree of noise contamination of a communications channel, it is possible to communicate discrete data (digital information) nearly error-free up to a computable maximum rate through the channel.[7]

The Shannon theorem states that given a noisy channel with channel capacity C and information transmitted at a rate R, then if $R < C$ there exist codes that allow the probability of error at the receiver to be made arbitrarily small. This means that, theoretically, it is possible to transmit information nearly without error at any rate below a limiting rate, C. The theorem does not address the rare situation in which rate and capacity are equal. The Shannon limit or Shannon capacity of a communications channel is the theoretical maximum information transfer rate of the channel, for a particular noise level.[8]

The channel capacity (i.e., the maximum data transfer rate of the transmission channel) can be determined from its bandwidth, by using the following formula for noiseless channel derived by Shannon:

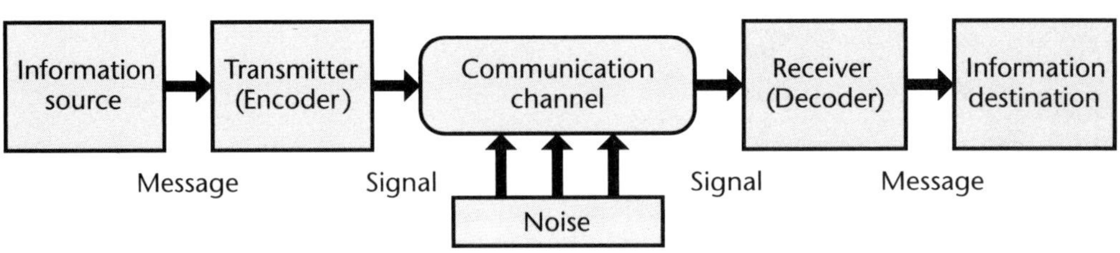

Figure 2.4 Noisy-channel block diagram.

7. This result was presented by Claude Shannon in 1948 and was based in part on earlier work and ideas of Harry Nyquist and Ralph Hartley. Shannon thought of both information sources and channels as random and used probability models for them. Shannon is widely known as the father of information theory.

8. Shannon only gave an outline of the proof. The first rigorous proof for the discrete case is due to MIT PhD student Amiel Feinstein in 1954.

$$C = 2B\log_2 M \tag{2.3}$$

where

C is the channel capacity in bits per second;

B is the bandwidth of the channel in hertz;

M is the number of discrete levels used to digitize an analog signal.

In the special case where only two levels, 'ON' and 'OFF' or 1 and 0 are used (binary), $M = 2$. Since, $C = 2B\log_2 2$, therefore $C = 2B$. As an example, the maximum data transfer rate for a voice channel of bandwidth 3,200 Hz carrying a binary signal would theoretically be 6,400 bps. In practice this figure is largely reduced by other factors (noise, for example) on the channel to approximately 4,800 bps.

In the presence of noise, the channel capacity will depend on the SNR and can be calculated using the following formula:

$$C = B\log_2\left(1 + \frac{S}{N}\right) \tag{2.4}$$

where

C is the channel capacity in bits per second, in the presence of noise;

B is the bandwidth of the channel in hertz;

S is the average received signal power over the bandwidth;

N is is the average noise or interference power over the bandwidth;

S/N is the SNR or the carrier-to-noise ratio (CNR) of the communication signal to the Gaussian noise interference expressed as a linear power ratio (not as logarithmic decibels).

We can say that faster data transmission requires channels that have a better SNR or wider bandwidth.

For practical purposes, when the SNR is very high, we can assume that $1+S/N$ is almost the same as S/N. In these cases, the theoretical channel capacity can be simplified to:

$$C = B \times \frac{S/N_{dB}}{3} \tag{2.5}$$

Problem:

Find the channel capacity of the typical analog telephone line, if the S/N is 30 dB, and the available bandwidth is 3 kHz.

Note: The value of $S/N = 1{,}000$ is equivalent to the S/N of 30 dB ($10\log 1{,}000 = 30$).

Solution:

$$C = B\log_2\left(1+\frac{S}{N}\right)$$
$$C = 3000\log_2(1+1000)$$
$$C = 3000\log_2(1001)$$
$$C = 3000\times 9.97$$
$$C \approx 30 \text{ kbps}$$

The channel capacity C can be calculated from the physical properties of a channel; for a band-limited channel with Gaussian noise, channel capacity is calculated using the Shannon–Hartley theorem.[9]

Problem:
We need to send 250 kbps over a noiseless channel with a bandwidth of 30 kHz. How many signal levels do we need? Discuss the solution.

Solution:

$$C = 2B\log_2 M$$
$$250{,}000 = 2\cdot 30{,}000\log_2 M$$
$$4.17 = \log_2 M$$
$$M = 2^{4.17}$$
$$M \approx 18$$

Discussion:
The number of levels is a power of two. $M_1 = 2^4 = 16$, and $M_2 = 2^5 = 32$, so for $M = 18$, we either have to reduce the bit rate or increase a number of levels. Because 16 levels is not sufficient for this bit rate, we will need at least 32 levels for this signal transmission.

2.3.4 Synchronous and Asynchronous Data Transmission

There are many serial data transmission protocols. The protocols for serial data transfer can be grouped into two types: synchronous and asynchronous.

For asynchronous data transmission, there is no common clock signal between the sender and receivers. Therefore, the sender and the receiver first need to agree on a data transfer speed. This speed usually does not change after the data transfer starts. Both the sender and receiver set up their own internal circuits to make sure that the data accessing follows that agreement. Although the difference in clocks is

9. Advanced techniques such as Reed–Solomon codes and, more recently, low-density parity-check (LDPC) codes and turbo codes, come much closer to reaching the theoretical Shannon limit, but at a cost of high computational complexity. Using these highly efficient codes and with the computing power in today's digital signal processors, it is now possible to reach very close to the Shannon's theoretical limit.

very small, it can accumulate fast and eventually cause errors in data transfer. This problem is solved by adding synchronization bits at the front, middle, or end of the data, so the receiver can correct the clock accumulation error.

The synchronization information may be added to every byte of data or to every frame of data and could account for up to 50% data transfer overhead and hence slows down the actual data transfer rate. It is good for data with large gaps—commonly used for communications over telephone lines.

In synchronous data transmission a continuous stream of data signals is accompanied by timing signals generated by an electronic clock to ensure that the transmitter and the receiver are in step (synchronized) with one another. A continual stream of data is then sent between the two nodes. Because there is no start and stop bits, the data transfer rate is quicker, although more errors will occur, as the clocks will eventually get out of sync, and the receiving device would have the wrong time that had been agreed in the protocol for sending/receiving data. As a result, some bytes could become corrupted (by losing bits). Ways to get around this problem include resynchronization of the clocks and the use of check bits to ensure the byte is correctly interpreted and received.

Synchronous data transmission is more efficient (lower overhead) than asynchronous data transmission, and most network protocols (such as Ethernet, SONET, and token ring) use synchronous data transmission.

2.3.5 Modulation

Modulation is the process of superimposing the modulating signal (information) on to a carrier, and this process can involve analog signals, digital signals, or both. The signal that carries information signal is called the carrier signal. On top of this signal there is superimposed information—the baseband signal. The process of combining these two signals is called the modulation [3].

2.3.5.1 Analog Modulation

A sinusoidal electromagnetic wave has three properties: amplitude, frequency, and phase. Phase and frequency are just different ways to view or measure the same signal change. Any one of these parameters can be modulated to convey information. In this case the analog carrier signal is modulated with an analog information signal. The three properties of a sinusoidal electromagnetic wave are described as follows:

- AM: The amplitude of the carrier is changed by the amplitude of the information signal. Frequency is kept constant. Sidebands are produced in the frequency domain at $f_c \pm f_M$. AM other than by a pure sine wave will cause additional sidebands normally at $f_c \pm nf_M$, where n = 1, 2, 3, 4, etc.
- FM: The frequency of the carrier is changed by the amplitude of the information signal. Amplitude is kept constant, as shown in Figure 2.5. An unmodulated RF signal in the time domain has only a single spectral line at the carrier frequency (f_c) in the frequency domain. If the signal is frequency modulated, the spectral line will correspondingly shift in the frequency domain.

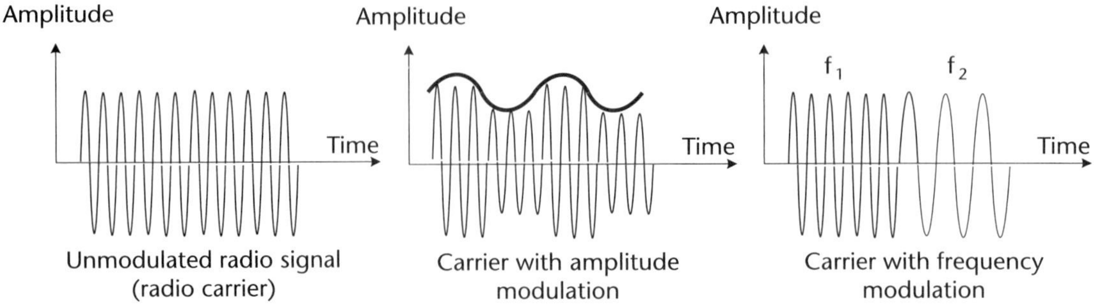

Figure 2.5 Analog modulation methods.

- Phase modulation (PM): The phase (time displacement) of the carrier signal is varied by the amplitude of the information signal.

To transmit useful information over the air, there are three main steps, briefly described as follows:

1. A pure carrier signal is generated at the transmitter.
2. The carrier signal is modulated with the information to be transmitted. Any reliably detectable change in signal characteristics can carry information.
3. At the receiver, the signal modifications or changes are detected and demodulated.

In AM radio each radio station is assigned 10 kHz to transmit a monochannel audio (band-limited to 5 kHz). AM is used to shift the baseband signal.

In FM radio each radio station is assigned 200 kHz to transmit a stereo audio. The left and right channels (each limited to 15 kHz) are multiplexed into a single baseband signal using AM. FM is used to shift the baseband signals.

2.3.5.2 Digital Modulation

Over the past 30 years, a major transition has occurred in communication systems, from simple analog AM and FM/PM to new, and more complex, digital modulation techniques [4]. Examples of digital modulation among others include the following:

- Amplitude shift keying (ASK): The amplitude of the carrier signal is varied (shifted) between two values (Figure 2.6). This is also known as on–off keying (OOK).
- Frequency shift keying (FSK): The frequency of the carrier signal is varied between two values by the modulating signal.
- Phase shift keying (PSK) or binary PSK (BPSK): The phase of the carrier signal is changed by the modulating signal. Depending on the number of discrete displacements, several bits of data can be transmitted simultaneously. For example, with four shift amounts (0 degrees, 90 degrees, 180 degrees, and 270 degrees), two bits (e.g., 00, 01,10 and 11) can be sent at a time.

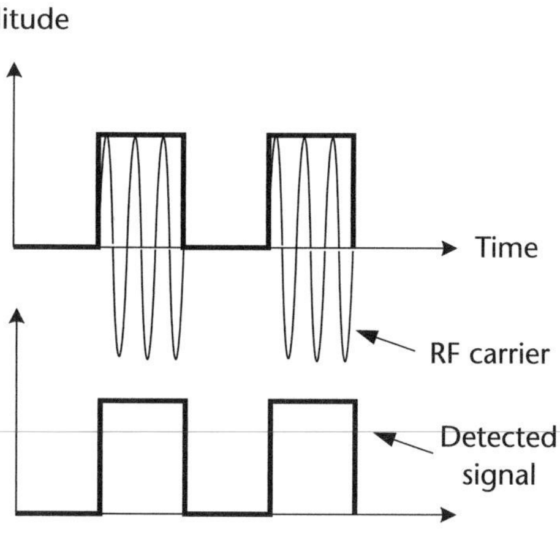

Figure 2.6 ASK.

- Quadrature amplitude modulation (QAM): Both the phase and the amplitude of the carrier are changed, making it possible to encode as many as four bits at a time.
- Trellis-coded modulation (TCM): Similar to QAM but includes extra bits for error correction.

The signal can be expressed in polar form as a magnitude and a phase, as shown by Figure 2.7. The carrier becomes a frequency and phase reference, and the signal is interpreted relative to the carrier. The magnitude could be either an absolute or relative value, and both are used in digital communications systems.

Amplitude and phase can be modulated simultaneously and separately, but this may be difficult to generate, and especially difficult to detect. Instead, in practical systems the signal is separated into another set of independent components: I (in-phase) and Q (quadrature). These components are orthogonal and therefore do not interfere with each other.

Polar diagrams are very common in many displays used in digital communications, although it is common to describe the signal vector by its rectangular

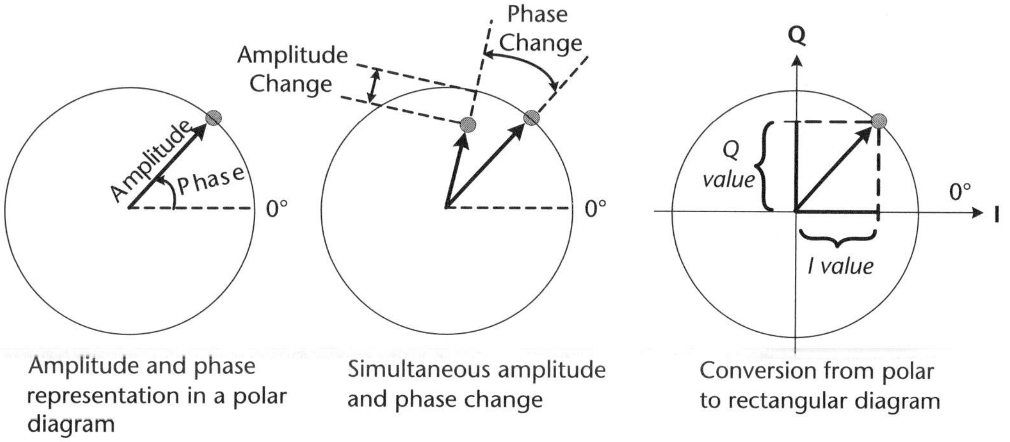

Figure 2.7 Signal changes in polar form.

coordinates of I and Q, a rectangular representation of the polar diagram. On a polar diagram, the I-axis lies on the zero-degree phase reference, and the Q-axis is rotated by 90 degrees. The signal vector's projection onto the I-axis is its "I" component and the projection onto the Q-axis is its "Q" component. I/Q diagrams are particularly useful because they mirror the way most digital communications signals are generated using I/Q modulator.

Signals that are orthogonal do not interfere with each other. They are two independent components of the signal. When recombined, they are summed to a composite output signal. There are two independent signals in I and Q that can be sent and received with simple circuits, thus simplifying the design of digital radios. The main advantage of I/Q modulation is the ease of combining independent signal components into a single composite signal and later splitting such a composite signal into its independent component parts.

2.3.5.3 Modulation Efficiency

Bandwidth efficiency describes how efficiently the allocated bandwidth is utilized or the ability of a modulation scheme to accommodate data, within a limited bandwidth. Table 2.1 shows the theoretical bandwidth efficiency limits for the main modulation types. Note that these figures cannot actually be achieved in practical radios since they require perfect modulators, demodulators, filters, and transmission paths.

In order to understand and compare different modulation format efficiencies, it is important to understand the difference between bit rate and symbol rate. The required signal bandwidth of the digital communication channel depends on the symbol rate, not on the bit rate.

$$\text{Symbol rate} = \frac{\text{Bit rate}}{\text{Number of bits per symbol}} \tag{2.6}$$

Symbol rate is sometimes called baud rate or modulation rate (note that baud rate is not the same as bit rate.) If more bits can be sent with each symbol, then the same amount of data can be sent in a narrower spectrum. This is why modulation formats that are more complex and use a higher number of states can send the same information over a narrower piece of the RF spectrum.

Table 2.1 Spectral Efficiency of Modulation Methods

Modulation Method	*Theoretical Bandwidth Efficiency Limits [bits/second/Hz]*
BPSK	1
QPSK	2
8PSK	3
16QAM	4

Let's say you have an eight-bit sampler, sampling at 10 kHz for voice. The bit rate, the basic bit stream rate, would be eight bits multiplied by 10k samples per second or 80 kbps.[10] Groups of k can then be combined to form new digits, or symbols, from a finite symbol set of $M=2^k$ such symbols. A system using a symbol set size of M is referred to as an M-ary system [5]. If one bit is transmitted per symbol, as with BPSK, then the symbol rate would be the same as the bit rate of 80 kbps. If two bits are transmitted per symbol, as in QPSK, then the symbol rate would be half of the bit rate or 40 kbps. QPSK is therefore a more bandwidth-efficient type of modulation than BPSK, potentially twice as efficient.

In 16-state quadrature amplitude modulation (16QAM), there are four I values and four Q values. This results in a total of 16 possible states for the signal. It can transition from any state to any other state at every symbol time. Since $16 = 2^4$, four bits per symbol can be sent. This consists of two bits for I and two bits for Q. The symbol rate is one fourth of the bit rate. So, this modulation format produces an even more spectrally efficient transmission; it is more efficient than BPSK, QPSK, or 8PSK. Note that QPSK is the same as 4QAM.

Due to their spectral efficiency, 32QAM, 128QAM, and 512QAM are also used for high-capacity microwave digital radios. Unfortunately, the symbols get closer and closer together and thus more prone to errors due to noise and distortion, as well as interference. For example, a signal using 32QAM is more susceptible to noise and distortion (and interference from other radio systems) than simpler QPSK.

2.3.6 Transmission of Analog and Digital Signals

Analog and digital signals can be transmitted using either analog or digital methods. (See Figure 2.8.) It is therefore possible to transmit the following:

- Analog signals using analog methods: This is typically the case with voice over a POTS telephone line. The PSTN has, for a long time, provided users

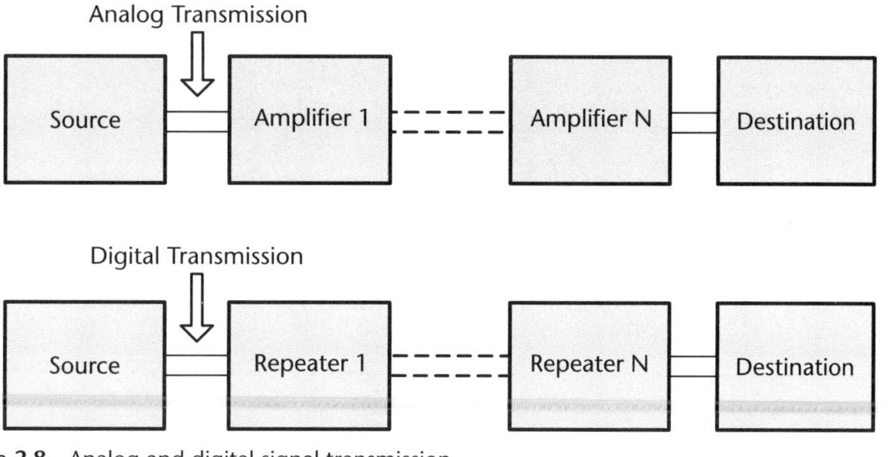

Figure 2.8 Analog and digital signal transmission.

10. For the moment we will ignore the extra bits required for synchronization, error correction, etc.

with dial-up telephone connections on demand. Each connection has supported analogue speech in the voice range of 300–3,400 Hz. For this purpose a normal dial-up or leased analog line is used, and no sophisticated conversion or modulation techniques are necessary.

- Analog signals using digital methods: The channel expects digital information, and, for this reason, the analog signal first has to be converted to a digital format. A typical solution is pulse code modulation (PCM); in the case of voice, the analog signal is typically sampled at 8,000 times per second and each sample is then converted to an eight-bit binary number giving a 64-kbps data rate.
- Digital signals using analog methods: In order for digital signals to be sent across a telephone connection, they must be converted to analog voice tones within the frequency range 300–3,400 Hz (this is done by a modem). A sine wave is modulated with the digital data, and, if necessary, several channels can be superimposed on the same physical circuit. Commercial solutions include modems for use on phone lines or one of the broadband alternatives.
- Digital signals using digital methods: This is an ideal situation, and no special conversion techniques are needed. The transmission channel could be any of the dial-up or leased line alternatives (T1/E1).

2.4 Data Networks

Data networks send data (information) from one electronic device to another. This complex process can be broken into discrete, sequential tasks. The sending device has to do the following:

1. Recognize the data.
2. Divide the data into smaller, manageable chunks (packets).
3. Add information to each chunk of data to determine the location of the data and to identify the receiver.
4. Add timing and error-checking information.
5. Put the data on the network and send it on its way.

Network client software operates at many different levels within the sending and receiving devices. Each of these levels (or tasks) is governed by one or more protocols.

2.4.1 About the OSI Layers

The OSI reference model is a description of layered communications and computer network protocol design. Basically, it divides network architecture into seven layers that, from top to bottom, are the application, presentation, session, transport, network, data link, and physical layers (Table 2.2). It is therefore often referred to as

Table 2.2 OSI Seven-Layer Model

Layer Number	*Layer Name*	*Layer Description*	*Protocols*
7	Application	Supports user applications, such as software for file transfers, database access, browsers and e-mail.	Telnet, SMTP, FTP, SNMP
6	Presentation	Responsible for converting protocols, translating the data, encrypting the data, changing or converting the character set, and expanding graphics commands.	HTTP, JPEG, ASCII, MP3, MPEG, encryption
5	Session	Allows two applications on different devices to open, use, and close a connection called a session. (A session is a highly structured dialog between two workstations.)	NetBIOS names, SQL, NFS
4	Transport	Ensures that packets are delivered error-free, in sequence, and without losses or duplications.	TCP, UDP, SPX
3	Network	Responsible for addressing messages and translating logical addresses and names into physical addresses and determines the route from the source to the destination device.	IP, IPX, DDP
2	Data Link	Sends data frames from the network layer to the physical layer. It controls the electrical impulses that enter and leave the network cable. The electrical representation of the data (bit patterns, encoding methods, and tokens) is known to this layer only.	IEEE 802.3, Frame Relay, ATM, PPP
1	Physical Layer	Transmits the unstructured, raw bit stream over a physical medium (such as the network cable). The physical layer is completely hardware-oriented and deals with all aspects of establishing and maintaining a physical link between communicating devices.	RJ-45, EIA-232, V.35, Ethernet

the OSI seven-layer model,[11] and most of the communication protocols used today have an OSI model–based layered structure.

Internetworking devices such as bridges, routers, and switches have traditionally been categorized by the OSI layer they operate at and the role they play in the topology of a network, described as follows:

- Layer 1: Hubs and repeaters/regenerators;
- Layer 2: Bridges and switches operate at layer 2 (data link layer);
- Layer 3: Routers operate at layer 3 (network layer).

Bridges and switches extend network capabilities by forwarding traffic among LANs and LAN segments with high throughput. Layer 2 refers to the layer in the communications protocol that contains the physical address of a client or server station. It is also called the data link layer or MAC layer. Layer 2 contains the address that is inspected by a bridge, switch, or PC NIC.

The Layer 2 address of every network device is unique, fixed in hardware by its manufacturer and usually never changed. Traditionally, products that were called switches operated by forwarding all traffic based on its layer 2 addresses.

11. The Open Systems Interconnection (OSI) reference model is a standardized seven layer model released in 1984, helping vendors to create interoperable network devices. The OSI model doesn't do any functions in the networking process; it is a conceptual framework for better understanding of complex interactions that are happening.

The spanning tree protocol, implemented in many layer 2 switches, prevents forwarding loops in switched networks. Unfortunately, this is achieved by shutting down redundant connections and never using them. In contrast, routers are able to keep redundant connections active and make use of this built-in redundancy to increase network reliability and performance. With layer 2 switching reaching the limits of its potential, the multilayer switch represents the next stage in the evolution of internetworking devices.

Multilayer switching is simply the combination of traditional layer 2 switching with layer 3 protocol routing in a single product, usually through a fast hardware implementation.

Routers perform route calculations based on layer 3 addresses and provide multiprotocol support and WAN access, but typically at the cost of higher latency and much more complex administration requirements. Layer 3, also called the network layer, refers to the layer in the communications protocol that contains the logical address of a client or server station.

Layer 3 contains the address (such as IP or IPX) that is inspected by a router that forwards the traffic through the network. The layer 3 address of a network device is a software setting established by the user network administrator that can and does change from time to time; only devices that need to be addressed by layer 3 protocols, such as IP, have layer 3 addresses. Traditionally, routers operated solely on layer 3 addresses.

TCP and IP are the best known of the Internet protocols and very often the term TCP/IP refers to the whole family of protocols. TCP/IP is a five-layer model where layers 1 and 2 are not actually defined since TCP/IP was designed to be independent of physical media.[12]

The enhanced performance architecture (EPA) model is a simpler reference model. In the OSI model, top three layers address the application message sent between stations, and the bottom four layers address the method used for the transportation of these messages. Some communication protocols use the EPA model, because not all features of the full seven-layer model are required and enhanced working of the remaining features is desired. The three-layer EPA model consists of only the physical Layer, the data link layer, and the application layer. DNP3 is an example of a protocol based on EPA model.

2.4.2 Intranets, Extranets, and VPNs

Intranets permit access by users who have access privileges to the internal LAN of the organization. Extranets are intranet-based but use extended, secure access to external users or enterprises. They both can use VPN to emulate the properties of a point-to-point private network (Figure 2.9).

VPN is a secure, private communication tunnel between two (or more) devices across a public network (like the Internet). These devices can be computers running some type of VPN software, or VPN-enabled routers. All of the VPN data is secure

12. The transmission control protocol/Internet protocol (TCP/IP) was developed in the 1970s by the U.S. Department of Defense to implement reliable network communications. The OSI model was developed after TCP/IP, but it attempted to maintain some similarities to the TCP/IP model.

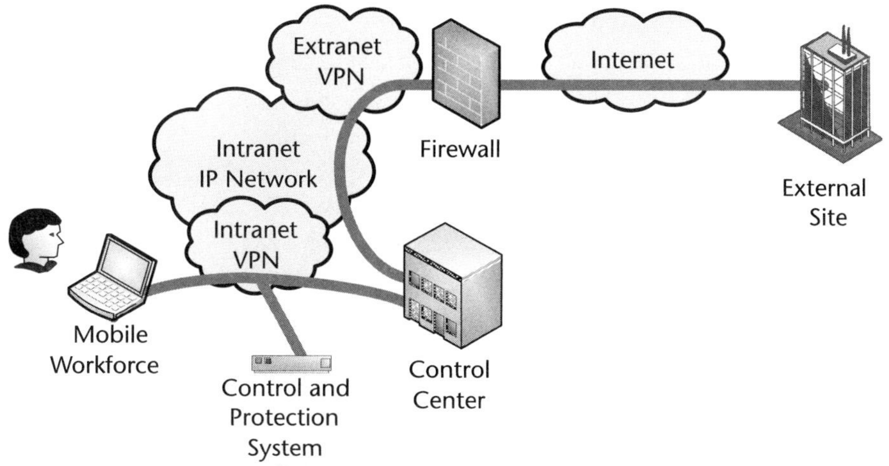

Figure 2.9 Intranet and extranet VPNs.

regardless of the fact it travels over a public network. VPN implementations use strong encryption and monitor traffic to ensure that no packets have been altered.

Early data networks allowed VPN-style remote connectivity through dial-up modems or through leased line connections utilizing frame relay and asynchronous transfer mode (ATM) virtual circuits, provisioned through a network owned and operated by telecommunications carriers. These networks are not considered true VPNs because they passively secure the data being transmitted by the creation of logical data streams.

They have been replaced by VPNs based on IP and IP/multiprotocol label-switching (MPLS) networks, due to significant cost-reductions and increased bandwidth provided by technologies such as digital subscriber line (DSL) and fiber-optic networks.

Tunneling is a method of transferring one network's data over another network [6]. The data (or payload) to be transferred can be in the form of the frames (or packets) of another protocol. Instead of sending a frame as it is produced by the originating node, the tunneling protocol encapsulates the frame in an additional header. The additional header provides routing information so that the encapsulated payload can travel over the intermediate network. The encapsulated packets are then routed between tunnel endpoints over the intermediate network.

The logical path through which the encapsulated packets travel through the intermediate network is called a tunnel. Once the encapsulated frames reach their destination on the internetwork, the frame is decapsulated and forwarded to its final destination. Tunneling includes this entire process of encapsulation, transmission, and decapsulation of packets.

There are two major protocols that VPN supports: point-to-point tunneling protocol (PPTP) and Internet protocol security (IPSec). Most broadband routers can pass PPTP traffic by forwarding port 1723, but IPSec is more complex. PPTP has decent encryption and also features "authentication" for verifying user IDs and passwords. IPSec is purely an encryption model and is much safer but does not include authentication routines.

2.4.3 Digital Bandwidth

Originally, the term bandwidth was used to specify the numerical difference between the upper and lower frequencies of a section of the electromagnetic spectrum. Analog bandwidth, measured in hertz, was explained in Section 2.2. On the other hand, bandwidth in the digital domain, or digital bandwidth, means the maximum achievable capacity of a digital transmission over an electronic pathway and is measured in bits per second.

How fast a file is transported from end to end depends on the size of the file but also on a digital bandwidth of the communication channel. The theoretical calculation for the transfer time of a file under ideal conditions is:

$$\text{Transfer time [seconds]} = \frac{\text{Bandwidth [bps]}}{\text{File size [bits]}} \tag{2.7}$$

2.5 Transmission Network Topologies

2.5.1 Multiplexing

Multiplexing is a technique by which information (voice or data) from more than one source is combined and transmitted using the same communications channel. Demultiplexing is an inverse operation in which we extract original signals. See Figure 2.10. Of course, traffic flows in both directions, so every multiplexer is also a de-multiplexer.

2.5.1.1 FDM

In FDM, multiplexing takes place in the frequency domain, and it represents the oldest multiplexing technique. Simply speaking, information is delivered using multiple frequencies.

FDM assigns a discrete (fixed) carrier frequency for each channel and then modulates each carrier with the input for the respective channel. Assuming that there are four channels and the total available bandwidth is 1 MHz, the maximum

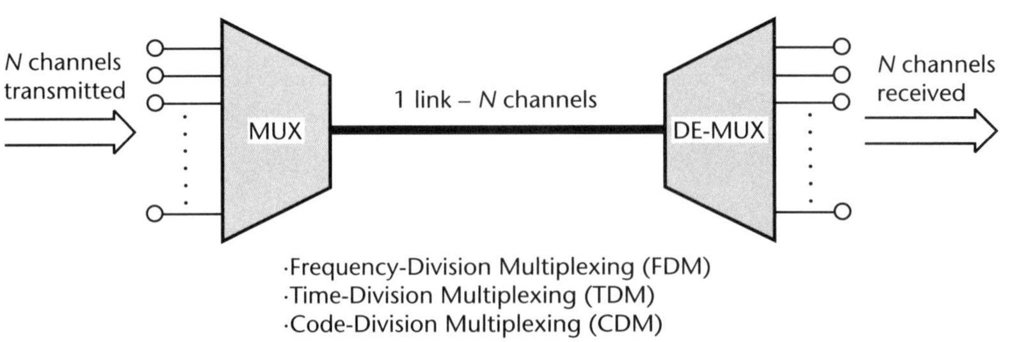

Figure 2.10 Multiplexing.

available bandwidth per channel is 0.25 MHz or 250 kHz. In practice this will be slightly less due to a guard band separating the channels in the frequency domain.

2.5.1.2 TDM

In TDM, multiplexing takes place in the time domain. Data streams from several sources are combined by sending small slices of data from each input across the common channel sequentially, in time, in a specific sequence. If there are four channels to be multiplexed like this, each source will have access to the channel for a quarter of the time.[13]

Typical examples of TDM are the T1 and T3 carrier systems in North America, where streams are created for each conversation or transmitted data and then combined with TDM on a common channel. TDM networks represent a reliable, well-developed technology with guaranteed and predictable service levels. Unfortunately, they are not suitable for the Ethernet transport and the introduction of new, bandwidth-intensive services.

2.5.1.3 Code-Division Multiplexing

Code-division multiple access (CDMA)[14] is a form of multiplexing that allows numerous signals to occupy a single transmission channel, optimizing the use of available bandwidth. CDMA employs ADC in combination with spread spectrum technology. The frequency of the transmitted signal varies according to a defined pattern (code), so it can be intercepted only by a receiver whose frequency response is programmed with the same code.

Examples are ultra-high-frequency (UHF) cellular telephone systems in the 800-MHz and 1.9-GHz bands.

2.5.2 T1 and E1

There are two standards for the first-order digital transmission systems. The T1 system, based on TDM principle and developed by Bell Laboratories, is used mainly in the United States, Canada, Taiwan, Jamaica, and a few other countries. North American T1 service providers often refer to the signal interfaces between the user and the network as DS1 signals. In the case of user-to-user interfaces, the term DSX-1 is used to describe those DS1 signals at the "cross-connect" point; therefore, DSX-1 is a physical interface of the T1 circuit.

Details of T1 systems can be found in literature [7] The following is a brief description (see Figure 2.11) of T1:

- There are 24 DS0 (64-kbps) time slots in a T1 line, providing a total bandwidth of 24×64 kbps = 1,536,000 bps (1,536 kbps).
- Another 8,000 bps are used as framing bits.

13. A special case of TDM is statistical multiplexing (STDM), where the input nodes that have nothing to send are skipped, thereby increasing the available time for the remaining nodes.
14. CDMA was originally developed over 30 years ago for military use.

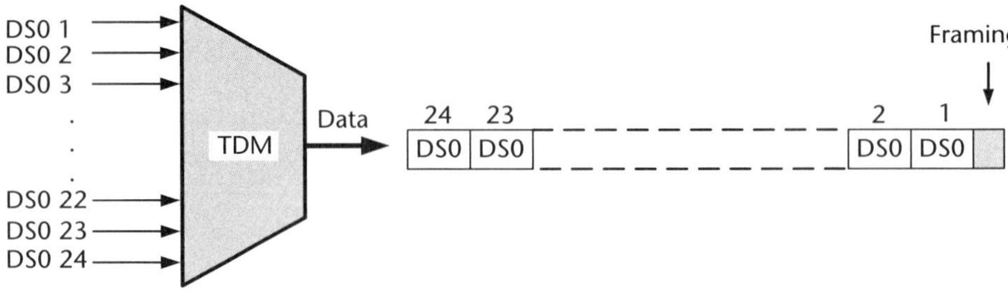

Figure 2.11 T1 structure.

- Adding up the total bandwidth of the 24 DS0 channels and the 8-kbps framing bits yields a 1.544-Mbps T1 data rate.

The North American digital hierarchy starts with a basic digital signal level of 64 kbps (DS0). Thereafter, all facility types are usually referred to as Tx, where x is the digital signal level within the hierarchy (e.g., T1 refers to the DS1 rate of 1.544 Mbps). (See Table 2.3.)

The basic format for T1 transmission facilities in Japan is similar to the North American ANSI standard and called J1 [8].

The ITU digital hierarchy's [9] basic level is the DS0 rate of 64 kbps.[15] These signals are usually delivered from the provider via twisted pair or coaxial cables. The following is a description of the CEPT (E1) digital hierarchy:

- There are 30 DS0 (64-kbps) time slots in an E1 line, providing a total bandwidth of 30×64 kbps = 1,920,000 bps (1,920 kbps).
- One 64-kbps time slot (TS0) is used for framing bits.
- Another 64-kbps time slot (TS16) is used for signaling of voice frequency channels.
- Adding up the total bandwidth of the 30 DS0 channels, framing, and signaling bits yields the 2.048-Mbps E1 data rate.

Table 2.3 North American Data Rates

Name	*Bit Rate (Mbps)*	*Number of Voice Channels*
DS0	64 kbps	1
DS1	1.544	24
DS1C	3.152	48
DS2	6.312	96
DS3	44.736	672
DS4	274.176	4,032

15. Most of the countries around the world use the E1 system defined by European Conference of Postal and Telecommunications Administration (CEPT).

2.5.3 Plesiochronous Digital Hierarchy (PDH), Synchronous Digital Hierarchy (SDH), Synchronous Optical Network (SONET), and ATM

Traditionally, transmission systems have been asynchronous, with each terminal in the network running on its own clock. In digital systems, clocking (timing) is one of the most important considerations. Asynchronous multiplexing uses multiple stages; lower-rate signals are multiplexed, and extra bits are added (bit-stuffing) to account for the variations of each individual stream and combined with other bits (framing bits) to form higher-level bit rates. Then bit-stuffing is used again to produce even higher bit rates. At the higher asynchronous rate, it is impossible to access these signals without multiplexing and demultiplexing [10].

PDH[16] signals have the essential characteristics of time scales or signals such that their corresponding significant instants occur at nominally the same rate. The PDH systems belong to the first generation of digital terrestrial telecommunication systems in commercial use.

In SDH, introduced in the late 1980s, the transmission is carried out in a synchronous mode, hence the name. The most important advantage in adopting this synchronous technology is to enable the mapping (dropping and inserting) of various user bit rates directly onto the main transmission signals, thus bypassing various stages of multiplexing and demultiplexing as was done in the case of earlier PDH technology [11]

North American SONET is a second-generation digital optical transport protocol. The fiber-based carrier network uses synchronous operations among such network components as multiplexers, terminals, and switches.

The number of new architectures and topologies are made possible as a result of this new technology. For instance, add/drop multiplexers (ADMs) have made it possible to use SONET/SDH terminals in a long chain with bit streams added or dropped along the way in a simple and effective manner [12]

By closing the chain at two ends, a ring configuration is possible, providing enhanced protection features [13,14] SONET-based rings create a robust, high-availability network that can "heal" itself automatically by routing around failures. The optical fiber rings offer security, high bandwidth, low signal distortion, and high reliability.

Common expectations for fiber and microwave elements are signal rates and interfaces, overhead processing, service channels, operations systems, and transmission quality. SONET/SDH microwave radios can easily integrate into a new or existing fiber-optic network. (See Figure 2.12.)

ATM is the complement of synchronous transfer mode (STM). STM is a circuit-switched networking mechanism whereby a connection is established between two termination points before data transfer commences and torn down when it is completed. In this way, the termination points allocate and reserve the connection bandwidth for the entire duration, even when not actually transmitting data.

ATM is an older transmission technology that uses fixed-size packets called cells. A cell is a 53-byte packet with five bytes of header/descriptor and 48 bytes of payload, or user traffic (voice, data, video, or their combination). It was developed to carry simultaneously both real-time traffic such as voice and high-resolution

16. The prefix plesio, which is of Greek origin, means "almost equal but not exactly," meaning that the higher levels in the ITU hierarchy are not an exact multiple of the lower level.

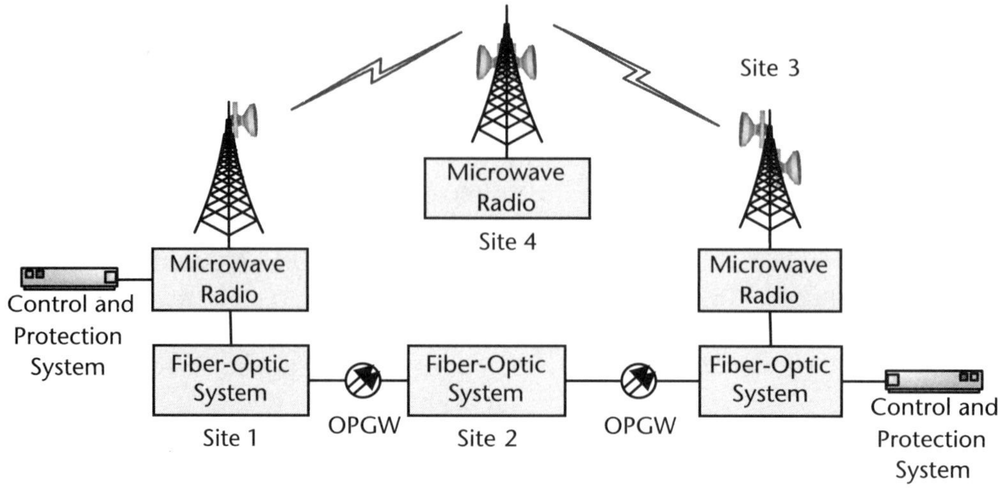

Figure 2.12 Hybrid microwave/fiber-optic ring.

video, which can tolerate some loss but not delay, as well as nonreal-time traffic such as computer data and file transfer, which may tolerate some delay, but not loss [15]

ATM traffic requires a very high-quality physical transmission medium with a good background error rate [also called residual BER (RBER)].[17] Microwave radio and fiber optics are ideal from this perspective, as they both offer RBER on the order of 10^{-13} [16].

2.5.4 Ethernet

Ethernet is the most popular physical layer LAN technology in use today. It defines the number of conductors that are required for a connection and the performance thresholds that can be expected and provides the framework for data transmission. A standard Ethernet network can transmit data at a rate up to 10 Mbps. Other LAN types include token ring, fast Ethernet, gigabit Ethernet, and 10-gigabit Ethernet.

Table 2.4 represents explanation of the terminology behind names; for example, 100Base-TX is a 100 Mbps over category 5 twisted pair. This technology constantly changes so new designations (and new standards) are added all the time (e.g., 10GE and 40GE).

Table 2.4 Ethernet Naming Convention

1,000	Network operates at 1,000 Mbps (1 Gbps)
Base	The type of signaling used is base
2 or 5	Maximum cable length in meters
T	Twisted pair cable
X	Full duplex capable cable
FX	Fiber-optic cable

17. RBER is caused by dribbling errors due only to hardware imperfections.

Packet-based equipment and early Ethernet did not have a mechanism to guarantee the delivery of message between points A and B; it worked under the "best effort" quality of service (QoS). Best-effort services such as e-mail are able to withstand the significant delay, packet reordering, and outages common on most IP networks. On the other hand, real-time services, such as voice-over-IP (VoIP), video and streaming media, and protection relaying in the electric utility environment demand a higher level of network performance with low latency and high network availability.

Today, new versions of Ethernet, including the MPLS technique, do have a QoS mechanism and settings that can guarantee the performance, in many ways mimicking the deterministic performance of SONET networks, while retaining the flexibility of packet-based networks. Different QoS requirements for voice, data, and video services must be supported by a well-designed IP- and Ethernet-based network.

Driven by the Metro Ethernet Forum (MEF) [17], a de facto standard "high-quality" transport service labeled carrier-grade Ethernet has been defined to meet these needs. Carrier Ethernet delivers the cost, scalability, and flexibility of Ethernet networks, but with TDM carrier-class reliability.

Networking equipment can waste lots of energy; energy-efficient Ethernet (EEE) is a set of enhancements that allow for less power consumption, accomplished by powering down switch ports when unused, adjusting transmission power based on cable length, and switching active switch ports into sleep mode when no traffic is expected, while retaining full compatibility with existing equipment. It was agreed that the EEE standard would deal with the "Base-T" interfaces that operate over twisted pair wiring because they comprise the vast majority of Ethernet deployments, especially at the edge of networks where the opportunities for energy savings are maximal. The standard also covers backplane Ethernet interfaces. The IEEE ratified the final IEEE 802.3az standard in September 2010 [18].

2.5.5 MPLS

MPLS is a technique (not a service) aimed to deliver anything from IP VPNs to metro Ethernet services, or even to provision optical services. The basic idea is to forward the packets based on a short, fixed-length identifier termed as a "label," instead of the network-layer address with a variable length match. The labels are assigned to the packets at the ingress node of an MPLS domain.

Inside the MPLS domain, the labels attached to packets are used to make forwarding decisions. Thus, MPLS uses indexing instead of a longest address match as in conventional IP routing. The labels are finally removed from the packets when they leave the MPLS domain at the egress nodes. By doing this, the efficiency of packet forwarding is greatly improved. Routers that support MPLS are known as label switching routers (LSRs).

This economical solution can be integrated seamlessly over any existing infrastructure, such as IP, frame relay, ATM, or Ethernet. Integration of MPLS

MPLS - MULTIPROTOCOL LABEL SWITCHING
VPN - VERTUAL PRIVATE NETWORK

application components, including layer 3 VPNs, layer 2 VPNs, traffic engineering, QoS, generalized MPLS (GMPLS)[18], and IPV6 enable the development of highly efficient, scalable, and secure networks.

An MPLS network has the ability to handle packets with particular characteristics (such as coming from particular ports or carrying traffic of particular application types) in a consistent fashion. Packets carrying real-time traffic, such as voice or video, can easily be mapped to low-latency routes across the network—something that's challenging with conventional routing.

The key architectural point with all this is that the labels provide a way to "attach" additional information to each packet—information above and beyond what the routers previously had.

MPLS was initially created to improve packet performance in the core of the networks but has commonly been adapted for other use cases, and one of the most important is traffic engineering. In MPLS traffic engineering, all configurations are done on a specific network node called the head end or ingress node. In MPLS traffic engineering[19], the traffic flow has bandwidth requirements, media requirements, and a priority versus other flows, among other characteristics.

In the past, in dealing with communications circuits, power utilities had the choice of owning or leasing TDM-dedicated hard-wired circuits. Today, they are facing the same dilemma, only with packet-based networks—whether to own or to lease. Although neither option sounds tempting to them, most power utilities will probably opt for building and owning their own private MPLS network over Ethernet, or at least initially upgrading parts of their existing networks. Of course, while they are migrating to new technologies, many will be using hybrid type networks—a mixture of legacy TDM and new packet-based networks.

2.5.6 Transmission Network Topology and Performance

The main objectives of the transmission network are to connect all the points of interest, satisfy the capacity demands, and provide reliable service using different media (microwave, copper, fiber optics, or satellites). During the transmission network build-out, it is imperative to establish a transmission network plan that will include all present traffic requirements as well as future expansion. With new smart grid requirements and multiple types of traffic potentially riding over the same communication channel, precise traffic calculations are becoming increasingly important.

2.5.6.1 Network Topology

Network topology is a logical model that describes how networks are structured or configured: daisy-chained (also called linear configuration), bus, star, ring, mesh, and hybrid (Figure 2.13).

18. GMPLS is a protocol suite extending MPLS to manage further classes of interfaces and switching technologies other than packet interfaces and switching, such as time division multiplexing, layer-2 switching, wavelength switching, and fiber-switching.

19. MPLS is an integration of Layer 2 and Layer 3 technologies. By making traditional Layer 2 features available to Layer 3, MPLS enables traffic engineering.

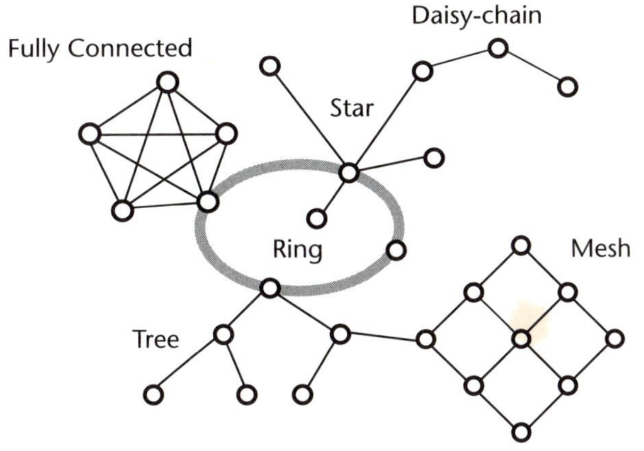

Figure 2.13 Communication network topologies.

Typically, long-haul (backhaul/backbone/WAN) networks will utilize some form of daisy-chained and/or ring topology. Networks inside substations will utilize mesh and hub/star topology, and networks in distribution systems will use some combination of daisy-chained, hub/star/mesh, and ring topology.

For a larger network, using a resilient ring configuration as a high-capacity backbone is recommended. The ring architecture is considered to be a reliable communication facility, as it provides automatic protection against the site hardware failures (batteries, towers, and antenna systems); radio, switch, or multiplexer (MUX) equipment failures; and radio propagation failures in the microwave network and cable cuts in the fiber-optic network.

Ring[20] architecture also provides basic user features such as simple operation, fault location, and maintenance. Ring configuration automatically provides alternative routing of traffic and no loss of traffic due to a single failure. For ultimate reliability, both directions can be hot standby (1+1)–hardware protected.

SONET/SDH technology have incorporated several protection/switching techniques from their inception, providing the ability for a network to detect the problem (under 10 ms) and heal itself automatically in the case of failure, with the restoration time less than 50 ms.

The important parameters that need to be considered when building digital communications systems include availability, latency (delay time), resynchronization time, and BER.

2.5.6.2 System Availability

For the communications network end users, one of the key factors to success is the ability to maintain a high standard of network performance. Five nines, the jargon term that means that a piece of equipment will function reliably 99.999 percent of the time (statistically, about five minutes of downtime per year), was used widely in the legacy TDM networks, but it is still in use today. Table 2.5 shows some of the commonly used percentage values for unavailability.

20. We should not mistake this type of ring with the token-ring in computer networks; they work on a completely different principle.

Table 2.5 Available and Unavailable Time per Year

Availability (% of time)	*Unavailability (% of time)*	*Unavailable Time (min/year)*
99.9999	0.0001	0.5256
99.9995	0.0005	2.628
99.9990	0.0010	5.256
99.9950	0.0050	26.28
99.9900	0.0100	52.56
99.9500	0.0500	262.8
99.9000	0.1000	525.6

The five nines requirement, correctly or incorrectly, has been applied to almost everything—from telecommunications equipment to microwave paths, fiber-optic links, and sometimes even to the all-inclusive end-to-end link. Unfortunately, to make things even more complicated, five nines will mean different thing to different people; so when we bring up the five nines requirement, it is advisable to have it precisely defined from the start.

Some of the main causes of IP network downtime are router hardware and software failures. In contrast to traditional central-office equipment, IP routers were not designed to support carrier-grade 99.999 percent availability; typical large IP networks achieve only between 99.95 and 99.99 percent availability. This is much more downtime than the five nines availability benchmark of legacy data networks.[21]

New generations of routing platforms have achieved 99.999 percent availability by providing full hardware and software redundancy in single router; in addition, hitless software upgrades are essential to the delivery of 99.999 percent availability because they eliminate the router downtime associated with software upgrades.

2.5.6.3 BER

In a digital communications system, information is transported encoded in ones and zeros, which correspond to the presence or absence of current. The communications channel and the associated equipment are not perfect, so at the receiver there will be some errors in that some "1's" will be detected as "0" and vice versa. The simple diagram in Figure 2.14 illustrates this concept.

BER[22] is used as a dimensionless measure of the performance of a communications link (i.e., radio path or optical fiber). The main reasons for the degradation of the communications link and the increase in corresponding BER are electronic

21. Studies have shown that link failures accounted for more than 30 percent of the outages in a large IP network. This is not particularly surprising, since IP networks traditionally have been built without locally protected links. Instead, these networks have relied on the ability of routers to route the traffic around failed links, producing unacceptable disruption to real-time and low-latency services.

22. In some texts, BER is referred to as the bit error rate instead of the bit error ratio. BER is an estimate formed by taking a ratio of bits in error to bits transmitted and it is dimensionless, so it is more accurate to use the word ratio instead of rate.

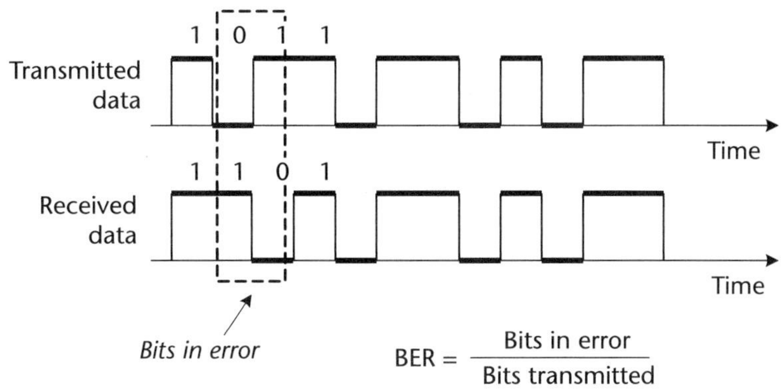

Figure 2.14 BER.

noise, hardware problems, and changes to the propagation path in the case of radio systems.

BER requirements are different for different services and systems; for example, radiolink has BER < 10^{-6}, while optical link has BER < 10^{-12}.

In practical tests, the BER is measured by transmitting a finite number of bits through the system and counting the number of bit errors received. By definition, BER is a quotient of bits received in error to the total number of bits received over a given period of time. For example, a BER of 10^{-10} means that there is one bit in error for every ten billion bits received.

In digital communication systems, BER is tested using predetermined stress patterns consisting of a sequence of logical ones and zeros generated by a test pattern generator. Test patterns emulating random data are called pseudorandom bit sequences (PRBSs)[23] and are based on standardized generation algorithms. They are called "pseudo" random because it is impossible to create a truly random signal using deterministic mathematical methods, but they are "sufficiently random" for intended purpose.

2.5.6.4 BER Versus Noise

Noise is an ever-present part of all communications systems. In analog systems, noise deteriorates the quality of the received signal, while in digital communications systems noise degrades the throughput because it requires retransmission of data packets or extra coding to recover the data in the presence of errors.

It's common to plot the BER in a digital communications system showing the average rate of errors for a given SNR. BER, noise, and digital modulation schemes are closely related parameters that define data transmission channel capabilities. Modulation schemes are compared through BER plots against E_b/N_0; see Figure 2.15(b), showing BER on the communications channel using different modulation schemes and in the presence of additive white Gaussian noise (AWGN).[24] For all

23. PRBS, also known as QRSS (Quasi Random Signal Source) tests are very useful to prove connectivity but cannot detect hidden or intermittent problems later. This is especially important in leased lines environment so end-to-end tests should also include stress patterns (All Ones, All Zeros, 3 IN 24, etc.).

24. In communications systems, the most common type of noise added over the channel is AWGN. It is additive because the received signal is equal to the transmitted signal plus the noise. It is white because it has

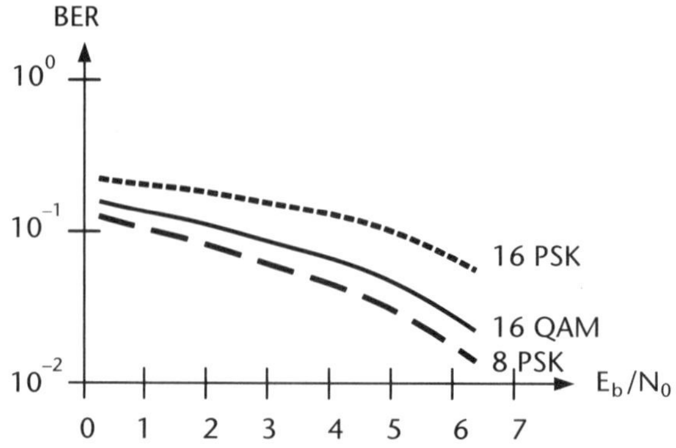

Figure 2.15 BER versus E_b/N_0 curves for different modulation schemes.

modulation schemes, the BER decreases monotically with an increase in E_b/N_0, which is the carrier power.

E_b/N_0 (the ratio of energy per bit to noise power spectral density) is an important parameter in digital communications. It is a normalized SNR measure, also known as the "S/N per bit" and allows for the comparison of the BER performance (effectiveness) of different digital modulation schemes. Because both factors are normalized, actual bandwidth is no longer important.

While higher order modulation rates are able to offer much faster data rates and higher levels of spectral efficiency for radio communications system, this comes at a price. The higher order modulation schemes are considerably less resilient to noise and interference.

2.5.6.5 Network Latency

Distances over which traffic and signaling are required vary considerably—from only a few tens of feet, where the devices of interest are located in the same substation, to applications on high-voltage lines where the distance between devices may be as much as a few hundreds of miles.

Signal propagation delay or latency describes the delay of a transmission from the time it enters the network until the time it leaves. Low latency is essential for real-time transmissions, including live voice conversations (but not voicemail messages, which are time-insensitive) and live two-way video (but not short video clips, which also are time-insensitive).

Latency is a phenomenon present in all the networks, terminals, and devices through which information may pass, plus the bottlenecks (and, therefore, delays) they may encounter. Delay can cause protocol time outs, retransmissions, and disruptions in data circuits and can inhibit voice transmissions. All types of transmission equipment, such as fiber-optic multiplexers, packet assemblers, microwave radios, and digital access cross-connects (DACS), add small amounts of delay as a result of their internal buffering. Satellite links add significant, and for some ap-

a constant power spectral density. It is Gaussian because its probability density function can be accurately modeled to behave like a Gaussian distribution.

plications unacceptable delay to a signal (for example, protection relays in electric utilities).

Power utility companies using communication systems to carry their SCADA signals require a very low delay, while protective relaying requires even lower delays. That is one reason why electric utility companies kept their old-fashioned analog microwave systems as long as they could—they introduce smaller delays than digital microwave systems and much smaller delays than fiber-optic systems.

Another parameter of interest is a variation of the delay when the traffic is rerouted over a different path. Utilities should carefully consider the latency, especially when it comes to use with protection relaying systems that perform time-sensitive information.

The main processing delay in a microwave radio is forward error correction (FEC) buffering. Total latency of the microwave link is a combination of the radio, free-space radio propagation, and multiplexer delays. Microwave radio manufacturers should be able to supply these numbers for their equipment.[25]

In utility communications networks, delay is expressed as a one-way end-to-end value.

2.5.6.6 QoS and Quality of (User) Experience (QoE)

The technical quality measures relating to data transport are commonly known as QoS. It is relatively easy to objectively measure the QoS within the network; in communications networks QoS can be defined in terms of blocking probability, network availability, setup time, network robustness, throughput, latency (delay), jitter, BER, lost packets and other measurable criteria.

The term QoE has been used to differentiate between QoS and user-perceived quality (usually audio and video). QoS will affect the QoE of the user, but there is a nonlinear relationship between the network parameters and the user perception of quality, so the traditional methodology of QoS is often ineffective in ensuring the desired user experience.

QoE (sometimes also abbreviated as QoX) is a relatively new term and may be defined as "the overall acceptability of an application or service, as perceived subjectively by the end user." It represents a purely subjective measure from the user's perspective of the overall value of the service provided. More information can be found in ITU-T E.800 [19].

2.5.7 Traffic Engineering

The term traffic engineering is widely used in the telephone voice world (landline and wireless). During the initial deployment of a new communications network, traffic needs to be estimated. After network deployment, the traffic is measured and analyzed. Then a statistical model is applied to the traffic pattern to make a prognosis and estimations. If the anticipated traffic pattern does not match well with the network resources, engineers need to remodel the traffic pattern in order to achieve a more optimum use of the resources.

25. Not discussed here are extremely low latency microwave radios, used only for very specialized applications like trade markets and financial institutions.

In the data communications world, traffic engineering provides an integrated approach to engineering traffic at layer 3 in the OSI model. The integrated approach means that routers are configured to divert from destination based forwarding to move the traffic load from congested parts of the network to noncongested parts. Traditionally, this diversion was done using overlay networks where routers use carefully engineered ATM or frame relay PVCs to distribute the traffic load on layer 2.

Traffic engineering is used to avoid inefficient mapping of traffic streams onto network resources. In such networks, one part of the network suffers from congestion during long periods of time, possibly continuously, while other parts of the network have spare capacity. A more efficient use of bandwidth resources allows for the avoidance of a situation where some parts of a network are congested, while other parts are underutilized.

Traffic engineering does not solve temporary network congestion caused by bursty traffic. This type of problem is better handled by an expansion of capacity or by classical techniques such as various queuing algorithms, rate limiting, and intelligent packet dropping.

2.5.8 Network Synchronization

Digital systems require all signals within a given transmission level to maintain a frequency relationship. If this relationship is not maintained, information will be lost, or transmission capacity will be underutilized. In addition, varying transmission times, inaccurate timing, and unstable network equipment can also cause bits traveling in a transmission path to arrive at a time that is in variance with their expected arrival. Solving timing-related problems is part of network synchronization.

Digital network connectivity depends on the availability of a reliable synchronization source to provide a timing reference to the network elements. Synchronization networks provide timing signals to all synchronization network elements at each node in a digital network. Buffer elements at important transmission interfaces absorb differences between the average local frequency and the actual short-term frequency of incoming signals, which may be affected by phase wander and jitter accumulated along the transmission paths [20].

2.5.8.1 Atomic Clocks and GPS

In telecommunications systems, stratum is used to describe the quality of a clock used for synchronization. The ANSI[26] Synchronization Interface Standard T1.101 defines profiles for clock accuracy at each stratum level, as does ITU standard G.810, and Telcordia/Bellcore standards GR-253 and GR-1244.

Stratum 1 is defined as a completely autonomous source of timing that has no other input other than perhaps a yearly calibration. The usual source of Stratum 1 timing is an atomic standard or reference oscillator. The minimum adjustable range and maximum drift is defined as a fractional frequency offset of 1×10^{-11} or less. Atomic standards, such as cesium clocks, use the natural oscillations of atoms in materials to define time and frequency standards.

26. ANSI-American National Standards Institute.

A Stratum 1 clock is an example of a primary reference source (PRS) as defined in ANSI/T1.101. Alternatively, a PRS source can be a clock system employing direct control from coordinated universal time (UTC) frequency and time services such as global positioning system (GPS) navigational systems. GPS may be used to provide high-accuracy, low-cost timing of Stratum 1 quality. Other Stratum levels represent clocks with lower accuracy. Any stratum clock will always control strata of lower-level clocks.

Inadequate timing may produce problems in any digital network, so the objectives have to be set very early in the planning process.

The GPS time standard is downlinked from the satellite as long as it is in range of the GPS receiver. However, there are intervals of time when a satellite is not in range (overhead), and no GPS time standard is available. This period without any GPS time update is called the GPS holdover time.

2.5.8.2 Interrange Instrumentation Group (IRIG)-B

IRIG[27] time codes are standard formats for transferring timing information. Atomic frequency standards and GPS receivers designed for precision timing are often equipped with an IRIG output. IRIG-B is a serial protocol and commonly used timing protocol for high-accuracy time synchronization for electric utility systems. IRIG-B is popular because it is fairly simple to understand, provision, and troubleshoot.

There are different versions of IRIG standards; the most commonly used of the standards is IRIG-B, then IRIG-A, then probably IRIG-G. For applications in power utilities, IEEE C.37.118 provides definitions for control function bits in an IRIG-B frame that conveys the time quality information of the present frame.

2.5.8.3 Precision Time Protocol (PTP)

IEEE 1588v2 (2008), also known as PTP, is a standard for providing precise timing and synchronization over packet-based, Ethernet networks.[28] Accuracy within the nanosecond range can be achieved with this protocol when using hardware-generated timestamps, although the protocol itself has no quality requirement. The transmission of the clock information over a packet network eliminates the need for alternative mechanisms, such as GPS or prohibitively expensive oscillators placed at the receiving nodes, and thus provides significant cost savings in network equipment as well as in ongoing installation and maintenance.

This synchronization solution transmits dedicated timing packets, which flow along the same paths with the data packets, reducing the cost of synchronization and simplifying implementation. Although IEEE 1588v2 systems add a small amount of additional traffic to the network load, they have several advantages; for

27. The standards were created by the TeleCommunications Working Group of the U.S. military's Inter-Range Instrumentation Group (IRIG), the standards body of the Range Commanders Council. Work on these standards started in October 1956, and the original standards were accepted in 1960. The latest version of the Standard is IRIG Standard 200-04.

28. An interesting new dilemma in the future may be how to handle many applications that depend on precision timing in computers and networks, which were designed to operate optimally without it.

example, they work in the data path, the most redundant and resilient part of the network, resulting in reliable operation. In addition, multiple transmission paths reduce redundant clock system costs.

The IEEE 1588v2 master function will typically be located at a central location, while the IEEE 1588v2 slave function will be integrated into the remote stations and receive the timing packets. The slave function consists of a clock recovery algorithm that works in tandem with the built-in high-quality oscillator.

It is important to note that IEEE 1588v2 protocol cannot improve the performance of synchronization or calibration through existing networks unless every network node is replaced with a node that supports an IEEE 1588v2 boundary clock or transparent clock [21].

2.6 Communication Networks and Systems in Substations

2.6.1 Modern Substation and EIA 61850

Substations are very important building block in the process of distributing electricity to homes and businesses in any country. The substation is where power is stepped down from high voltages used in transmission to the lower voltages used in homes and businesses. They also contain switches, capacitors, transformers, and other devices that utilities use to manipulate the grid to keep load flows balanced and routed to the correct locations. Substations are central to everything the electric grid does, and therefore central to any future smart grid.

Switches and capacitor taps have to be adjusted manually, and someone has to drive to the substation to make the necessary changes. With modern AMI[29], home automation (HA), and DA, some of those changes can now be accomplished remotely. Computer systems can now read inputs from substations and provide displays in utility control stations. It is important to note that security and access control are becoming increasingly important these days and represent a strict requirement for electric utilities [22].

To get to a smart grid, utilities are starting to overlay communications systems, many of them IP-based, over the network of substations (Figure 2.16). The process bus connects plant equipment (intelligent switch equipment as well as current and voltage transformers) and IEDs. The traffic mainly comprises SV, GOOSE and MMS protocol messages. The process bus must provide guaranteed QoS for real-time SV and GOOSE traffic and reliable delivery for MMS traffic.

The station bus interconnects the entire substation and provides connectivity between central management and the individual bays. The station bus also connects the devices within a bay and between different bays. The traffic is mainly GOOSE, MMS, simple network management protocol (SNMP) and file transfer protocol (FTP) messages as well as video and VoIP. SV messages are also sometimes used in busbar protection applications.

Traffic generated by multiple applications—for example, teleprotection, SCADA and SV—is carried over the substation communications infrastructure. To

29. AMI is a connectivity solution for smart energy grids and it is used for transporting (backhaul) meter data and meter management traffic between collector points and control rooms. Different media can be used for AMI; for example - radio, WiMAX, or BPL/PLC field area network (FAN) are utilized.

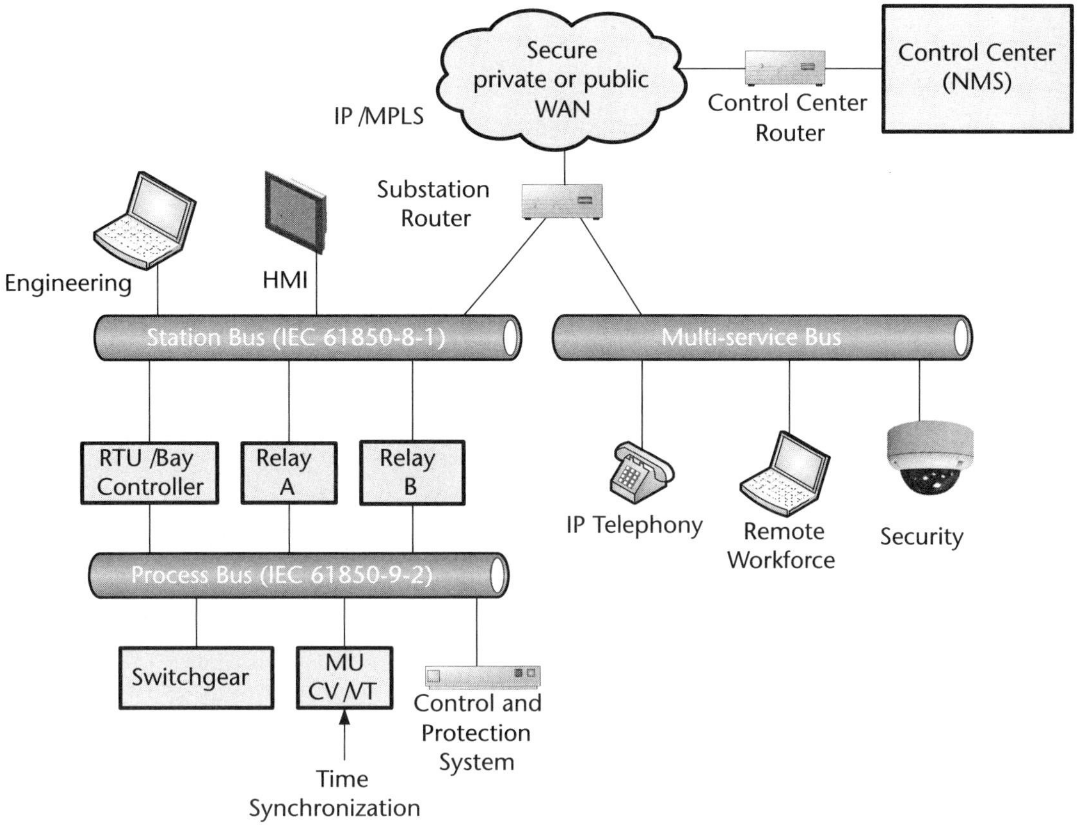

Figure 2.16 Substation communications.

support a multiservice architecture, VLAN bridging is required to isolate each application. Each virtual bridge must have its own MAC forwarding table to process the traffic, in addition to a QoS policy and bandwidth partitions. There are two Ethernet networking options: virtual private LAN service (VPLS) over MPLS ring or VLAN ring.

SCADA provides voltage and current measurements at critical grid nodes every two to four seconds. SCADA systems require minimal latency (typically less than 100 ms) in order to optimize polling performance and prevent communications “front ends” from timing out. While bandwidth requirements for SCADA-like operations may grow with smart grid deployments, traditional SCADA systems are not bandwidth-intensive.[23].

There are over 3,000 electric utilities across the United States, with different topographies, regulatory regimes, and legacy communications systems. Today, electric utilities use a variety of communications networks—wired and wireless, licensed and license-exempt, private and public (leased) and fixed and mobile, narrowband and broadband. In addition, these utilities use different protocols. Some of them will be described in the following sections of this chapter.

The hardwiring of traditional substations required several miles of secondary wiring in ducts and on cable trays. This not only increased the cost but also made the design inflexible. Ethernet is slowly finding its way into the electric utility space, and IEC 61850 Communications and Systems in Substations[30] is a new international standard comprised of 10 parts that outline a complete framework for substation automation including EMI immunity and environmental requirements for communications networks and systems in substations. Interdevice communications are through Ethernet and use the same communication protocol, IEC 61850, and both the cost and physical footprint of the substation have been reduced [24].

Ethernet-based IEC 61850 enables integration of all protection, control, measurement and monitoring functions within a substation and provides the means for high-speed substation protection applications, interlocking, and intertripping. The abstract data models defined in IEC 61850 can be mapped to a number of protocols that can run over TCP/IP networks or substation LANs using high-speed switched Ethernet to obtain the necessary response times, including below 4 ms for protective relaying.

2.6.2 Communication Protocols and Standardization Challenges

2.6.2.1 About Protocols

Communication systems use well-defined formats (protocols) for exchanging messages within or between electronic devices, where each message has an exact meaning intended to provoke a particular response of the receiver, and they can therefore be implemented as hardware, software, or both. Communications protocols have to be agreed upon in advance by the parties involved.

The electric utility industry has gone through many phases with protocols for control systems—in the beginning, practically no one considered interoperability. In the 1980s, there were only a few shared and another few proprietary protocols along with a few "utility-centric" protocols.

To reach an agreement, a protocol may have to be developed into a technical standard. In the late 1980s, the IEEE PES Substations Committee formed a working group (WG) to investigate the problem of proprietary protocols and to determine a reasonable solution.[31] Two protocols were selected: distributed network protocol, version 3, (DNP3) and IEC 60870-5.

There are two other IEC activities that are sometimes mentioned in relationship to the SG: the Common Information Model (CIM) IEC 61968 and 61970 models and IEC 61850 Substation Communication Protocol. The CIM models define the basic elements of the grid and their interconnection to the geographic information system (GIS).

The IEC 61850 protocol standard for substations enables the integration of all protection, control, measurement, and monitoring functions by one common protocol. It includes a number of features that should be considered for any control

---

30. Multiple protocols exist for substation automation, which include many proprietary protocols with custom communication links. Interoperation of devices from different vendors would be an advantage to users of substation automation devices. An IEC project group of about 60 members from different countries worked in three IEC working groups from 1995. They responded to all the concerns and objectives and created IEC 61850.

31. In 1997 IEEE Standard 1379 "IEEE Recommended Practice for Data Communications between Intelligent Electronic Devices and Remote Terminal Units in a Substation" was adopted and published. This standard is obsolete today.

system, including the object definitions and concepts, the use of XML files for defining IEDs and master station data bases, and naming conventions.

Dozens of serial data interfaces are still used today; some of the oldest legacy interfaces still in use are EIA-232, EIA-422, and EIA-485. The whole purpose of a serial interface is to provide a single path for data transmission over a cable or by using radio technology. Serial interfaces can be used to provide standardized logic levels from transmitters to receivers, define the transmission medium and connectors, and specify timing and data rates. In some cases, they can perform serial-to-parallel and parallel-to-serial conversion or specify a basic data protocol.

New, Ethernet-based protocols are increasingly common and so are conversions between different protocols and different types of media.

2.6.2.2 EIA-232

EIA-232 (also known as RS-232)[32] was originally established in 1962 as a method of connecting data terminal equipment (DTE) such as electromechanical teletypewriters to data communications equipment (DCE). Over the years its use has included connections to video terminals, computers, and modems. The first personal computers included an EIA-232 called a serial port for connection to a printer or other peripheral device.

Today, it is still widely used in embedded computer development systems, scientific instruments, and all sorts of industrial control equipment. Generally speaking, the EIA-232 is best for short-distance low-speed applications.

A physical connector for EIA-232 is not defined, and although DB-9 is quite common, devices come equipped with many different types of connectors.

2.6.2.3 EIA-422

EIA-422 is a technical standard that specifies the electrical characteristics of a digital signaling circuit. Differential signaling can transmit data at rates as high as 10 Mbps or may be sent on cables as long as 4,000 ft. Some systems directly interconnect using EIA-422 signals, or EIA-422 converters may be used to extend the range of EIA-232 connections. The standard only defines signal levels; other properties of a serial interface, such as electrical connectors and pin wiring, are set by other standards.

EIA-422 is also specified for multidrop applications where only one transmitter is connected to, and transmits on, a bus of up to 10 receivers.

2.6.2.4 EIA-485

EIA-485 is a superset of EIA-422 and expands on the capabilities. It defines not only a single device-to-device interface but also a communications bus that can be used to form simple networks of multiple devices, allowing up to 32 devices to

32. The official name of the standard is Electronic Industries Association / Telecommunications Industries Association EIA/TIA-232-F. The letter F designates the most recent standard modification and update. The standard is essentially the same as the International Telecommunications Union – Telecommunications (ITU-T) specifications V.24 and V.28.

communicate through the same data line. Its configuration and specifications also extend the range and data rate beyond the EIA-232 interface capabilities.

See Table 2.6 for a comparison of the main parameters among serial interfaces. RS485 is most popular in applications where noise immunity and long-distance capability are required.

2.6.2.5 Harris-5000/6000

Developed by Harris (GE Harris), Harris-5000/6000 supports communication between a SCADA system and RTUs of the same denomination. It has been widely used in North America and comes in several variants: Harris-5000, 5500, and 6000. The main difference between these versions is the extended address range in Harris-6000.

Harris-5000/6000 operates in half-duplex mode based on the master/slave principle. One master can communicate with up to 63 RTUs, while each RTU only responds to explicit requests from the master. Harris-5000/6000 does not support quality identifiers, while transmission of time stamps is however possible.

2.6.2.6 V.35

V.35 is a standard for synchronous serial interfaces, commonly employed on wide area networking and telecommunications equipment. It supports higher data transfer rates and connectivity between DCE and DTE over digital lines, with a maximum speed of 2 Mbps. The physical and electrical characteristics of this interface are now specified in ITU-T recommendations V.10 and V.11.

V.35 is a combination of balanced (like EIA-422) and common Earth signal interfaces (like EIA-232). The control signals in V.35 are common Earth single wire interfaces because these signal levels are mostly constant or vary at low frequencies. The high-frequency data and clock signals are carried by balanced lines.

The large 34-pin connectors are much more expensive than DB25, as well as special nonstandard cables required, making everything to do with V.35 more expensive.

2.6.2.7 "Ear and Mouth" (E&M)

E&M is a type of supervisory line signaling that uses DC signals on separate leads, called the E lead and M lead. The E&M analog interface is typically found in PBXs (4W analog E&M interface). E&M was originally developed to allow PABXs in

Table 2.6 Comparison of EIA-232, EIA-422, and EIA-485

Specifications	*EIA-232*	*EIA-422*	*EIA-485*
Mode of operation	Single-ended	Differential	Differential
Total number of drivers and receivers on one line (one driver active at a time for EIA-485 networks)	1 Driver 1 Receiver	1 Driver 10 Receiver	32 Drivers 32 Receivers
Maximum cable length	50 ft (2,500 pF)	4,000 ft	4,000 ft
Maximum data rate (40 – 4,000 ft for EIA-422/EIA-485)	160 Kbps (and up to 1 Mbps)	10 Mbps	10 Mbps

different geographic locations to communicate over an analog private circuit. There are six different types of E&M signaling.

2.6.2.8 ITU G.703

The ITU-standard G.703 describes the physical and electrical characteristics of hierarchical digital interfaces. The principles in G.703 are applicable to interfaces with data rates ranging from 64 kbps to 2.048 Mbps.

Under G.703 data can travel over balanced (120-ohm) pairs or unbalanced (dual 75-ohm) coax wires; RJ-45 connectors (or just open terminals) are used for the balanced version, and a pair of BNC connectors is used for the unbalanced interface.

2.6.2.9 IEEE C37.94

The IEEE C37.94 (published in 2002) is a standard for N-times 64 kbps optical fiber interfaces between teleprotection and multiplexer equipment; it provides interchangeable communications between different manufacturers' teleprotection and multiplexer devices using multimode optical fiber. (See Figure 2.16.) It is used for short multimode fiber-optic communication links of up to 2 km (1.25 mi).[33]

The standard defines clock recovery, jitter tolerances, physical connection methods, and the equipment failure actions for all communications link failures.

2.6.2.10 Distributed Network Protocol, version 3 (DNP3)

DNP3 was released in 1993 as a set of communications protocols used between components in process automation systems and probably is the most widely deployed and used protocol in North America. Its main use is in electric and water utilities.

Based on IEC 60870-5, it is used for inside substations and for substation to master station (or control center) communications. It plays a crucial role in SCADA systems, where it is primarily used for communications between a master station and RTUs or IEDs.

The ownership and maintenance of the DNP3 protocol has been under the control of the DNP Users Group, an open-membership not-for-profit corporation since 1996. The enhancements recommended for the protocol by the technical committee and approved by the membership have led to its wide-scale acceptance and enhanced functions.

Cybersecurity features developed by the IEC, Technical Committee 57 (TC57) WG IEC 62351-5 have been added to the DNP3 protocol.

The IEEE adopted DNP3 as IEEE Std. 1815-2010 in 2010. In April of 2012, the IEEE Std. 1815-2012 was approved for publication. The 2012 version of the standard includes features for Secure Authentication Version 5 (SAv5). The new

33. Although not covered in the specifications, some vendors offer a singlemode fiber version of IEEE Std C37.94-2002. The user should verify compatibility of different equipment singlemode interfaces for proper operation.

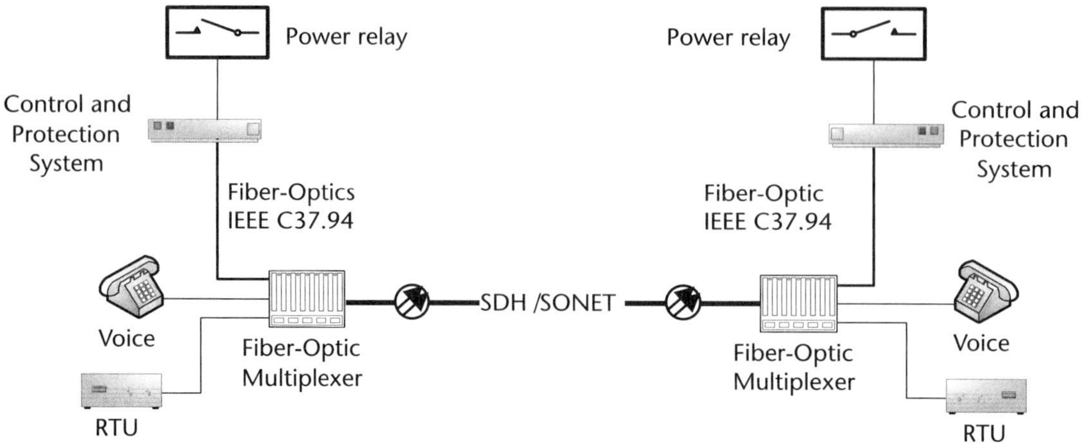

Figure 2.17 IEEE C37.94.

version is capable of using public key infrastructure (PKI)[34], and it facilitates remote key changes.[35]

2.6.2.11 Modbus

Modbus is an application layer messaging protocol—positioned at level 7 of the OSI model—that provides client/server communication between devices connected on different types of buses or networks. Modbus[36] was defined in 1979 to interface with PLCs. It is used to read and write registers and coils and has become a standard in many industries and utilities that is still used extensively today. Modbus is often used to connect a supervisory computer with a RTU in SCADA systems.

Originally designed for PLCs of the time, it is a request/reply protocol (also called master-slave) that allows a master to read or write bits in registers of the slave devices. It can be implemented on simple serial connections, most commonly over EIA-485 (also over EIA-232 or EIA-422), and it has lately been migrated to Ethernet (Modbus-IP).

Modbus is robust, simple to use and implement, and fairly secure and has very widespread usage, and fairly secure. Different implementations use wireline, radio communication, such as in the ISM license-exempt band, and even short message service (SMS). One of the more common designs of radio networks makes use of mesh networking.

Modbus has some drawbacks, for example, there is no time stamp or quality on data, no standard for data formatting, no event data, and high latency and timing issues. For more information about MODBUS, see [25].

34 Although not covered in the specifications, some vendors offer a singlemode fiber version of IEEE Std C37.94-2002. The user should verify compatibility of different equipment's singlemode interfaces for proper operation.

35. This approach is not based on encryption; rather it is based on verifying the sender of the message with the assurance that the message has not been tampered with. An attacker can see the message since it is not encrypted, but cannot tamper with it or send unauthorized messages without the key.

36. Modbus is a serial communications protocol originally published by Modicon (now Schneider Electric).

2.6.2.12 Generic Object-Oriented Substation Event (GOOSE)

GOOSE is probably one of the most important features of IEC 61850 standard. GOOSE[37] is a mechanism for the fast transmission of time-critical substation events, such as commands, alarms, indications, and messages. A single GOOSE message sent by an IED can be received and used by several receivers.

GOOSE takes advantage of the powerful Ethernet and supports real-time behavior (Figure 2.18). One device (sender) publishes information; only the subscriber devices receive it. Here, Ethernet replaces the point-to-point hardwired copper connection between different IEDs. Since GOOSE does not use a handshake mechanism, speed performance substation automation functions is improved. To ensure the highest level of reliability, GOOSE messages are repeated as long as the state persists.

GOOSE is fast enough for control schemes (4 ms), so it can be used for tripping switchgear, starting the disturbance recorder, and providing position indication for interlocking. Some North American utilities are using DNP3, Modbus, and IEC 61850 GOOSE messages on the same substation LAN. Simply speaking, GOOSE acts as a rescue manager for the devices configured for this service. Whenever there is a fault—whether due to an overvoltage problem, current leaking situation, or alarm trip fault—this device will act as a manager to a specific device facing such a fault. GOOSE actually has the ability to break a circuit switch, but only if configured for that functionality.

2.6.2.13 The Inter-Control Center Communications Protocol (ICCP)

ICCP[38] is being specified by utility organizations throughout the world to provide data exchange over WANs between utility control centers, utilities, power pools, regional control centers, and nonutility generators.

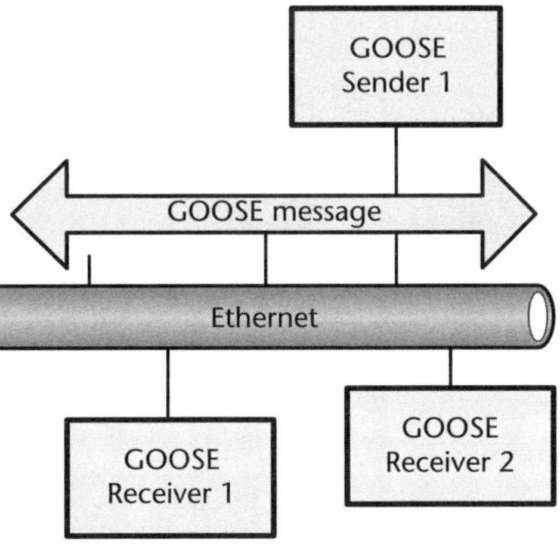

Figure 2.18 GOOSE block diagram

37. It is adopted as IEC 61850 from UCA2 GSE.

38. ICCP is also known as IEC 60870-6/TASE.2.

The real-time, interutility exchange of data has become critical to the operation of interconnected systems. Regional independent system operators (ISOs) and transmission operators with connections to various transmission and distribution utilities depend on the real-time exchange of data with all grid and market participants. The ICCP protocol is an internationally recognized standard for communications between electric utility control centers. ICCP uses the manufacturing messaging specification (MMS) for the messaging service needed by ICCP.

At the top level there is typically a system operator with coordination responsibilities for dispatch and overall system security. Below this are regional transmission companies that tie together distribution companies and generating companies. ICCP allows the exchange of real-time and historical power system information including status and control data, measured values, scheduling data, energy accounting data and operator messages.

ICCP does not provide authentication or encryption; they are normally provided by lower protocol layers.

2.6.2.14 Wireless Highway Addressable Remote Transducer Protocol (HART)

WirelessHART is an industrial wireless sensor networking technology based on the HART.[39] HART is a bidirectional communications protocol that provides data access between intelligent field instruments and host systems. A host can be any software application from technician's handheld device or laptop to a plant's process control, asset management, safety or other system using any control platform.

Developed as a multivendor, interoperable radio standard, WirelessHART was defined for the requirements of process field device networks. The protocol utilizes a time synchronized, self-organizing, and self-healing mesh architecture. The protocol supports operation in the 2.4-GHz ISM band using IEEE 802.15.4 standard radios [26].

In 2010, WirelessHart was approved by the International Electrotechnical Commission (IEC), making it the first wireless international standard as IEC 62591.

2.6.3 Interfacing Legacy Equipment

As we discussed in Section 2.5.1, multiplexers are devices that combine different services (applications) on the same communications circuit. Presently, substation multiplexers support numerous communications services, including SCADA and telemetry connectivity for analog/digital RTUs, teleprotection, analog voice and VoIP, PMRs and trunked radio systems, hotline/teleconferencing and PBX extension, controllers and relays, intersubstation LAN and NOC connectivity, PLC, and CCTV [27].

There are two basic approaches used by utilities today. One is to have new services over existing (legacy) networks or to have legacy services over new networks.

39. In 2012, the HART Communication Foundation officially added a HART over Internet protocol (HART-IP) capability to the HART Protocol, a specification for transmitting the HART Protocol over the increasingly common IP using either Transmission Communication Protocol (TCP) or User Datagram Protocol (UDP). HART-IP is an application protocol and defines rules for organizing and interpreting data and is essentially a messaging structure that is independent of the underlying physical layer.

Often, utilities will have a combination of these two network approaches implemented in their organization.

Utilities have many pieces of equipment with old-fashioned (legacy) serial interfaces (e.g., RS232 and RS422) and serial/Ethernet protocol converters[40] are sometimes required. (See Figure 2.19.) Some commonly used media converters are copper-to-fiber-optic media converters, fiber-fiber converters, and fiber-Ethernet converters.

2.6.4 Meter Data Management (MDM)

Home energy management (HEM) systems allow every point of the smart home to be connected and controlled from a central point, enabling power efficiency and energy optimization. Key features of HEM are listed as follows:

- They collect real-time energy consumption from smart meter and power consumption data.
- They control activation/deactivation of home appliances.
- They generate a dashboard to provide feedback about power usage.
- They provide control menus to control appliances.
- They provide a link to the utility and/or Internet.

From HEMs, data is sent to utilities for further processing. Utilities installing AMI can improve efficiency through MDM, which can play a crucial role in AMI rollouts, billing, and change management.

MDM must include the ability to validate, estimate and edit (VEE) AMI data so that clean and complete data can be delivered to utility systems. In the context

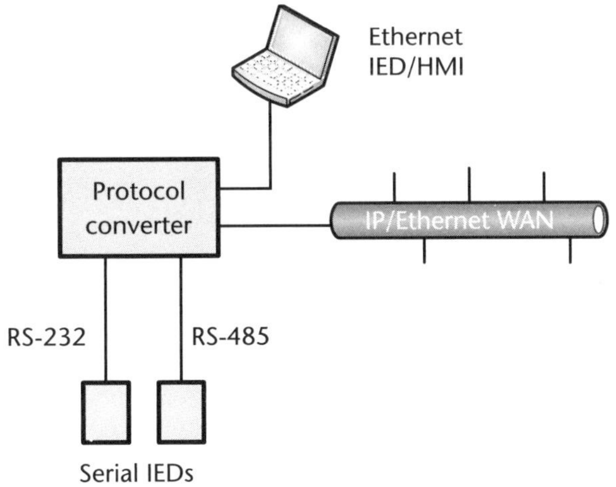

Figure 2.19 Protocol converters.

40. Protocol converters could add a significant amount of latency, so delay-sensitive applications (like protection relays, synchrophasors, etc.) may not be a good application for them. In addition, some legacy devices use protocols that can be very difficult to encapsulate into IP packets. If these circuits are using leased analog lines, they will have to be upgraded before year 2020 at which time most telecom carriers will stop offering TDM circuits.

of AMI interval data, VEE verifies the accuracy and completeness of the meter data and then fills in the gaps. Because utilities and their employees get better information sooner, they can work more efficiently, make fewer mistakes, and make better informed decisions. As an added benefit, their customers can view information online the same day it is received.

2.7 Review Questions and Problems

1. An older computer base system is "octal" or base eight. The digits 0, 1, 2, 3, 4, 5, 6, and 7 are used to represent any number. For example, $125_{10} = 1 \times 8^2 + 7 \times 8^1$ x $5 \times 8^0 = 175_8$. Write number 66 using number system with the base eight. (Solution: $66_{10} = 102_8$)
2. List countries (outside the United States and Canada) that use T1 digital hierarchy.
3. Is it possible to mix communications equipment having T1 and E1 interfaces? Under what circumstances can E1-based (CEPT) equipment be used in North America?
4. Describe terms like performance, availability, and quality in the context of communications networks.
5. What are the advantages of SONET over older PDH transmission systems?
6. Why is SNR important in communications systems and used as one of the main design parameters?
7. Find the channel capacity of the typical analog circuit, if the SNR is 28 dB and the available bandwidth is 4 kHz. (Solution: 26.1 kbps)
8. Amplitude modulation changes ______ of the carrier signal while frequency modulation changes _______ of the carrier signal.
9. What is the BER in a communication systems that detects 10 errored bits every hundred billion bits? (Solution: BER = 10^{-10})
10. What is the annual availability of the communications system that is 24 minutes per year out of service (i.e., unavailable)? (Solution: Availability = 99.99543 % of time)

References

[1] Lehpamer, H., *Transmission Systems Design Handbook for Wireless Networks,* Norwood, MA: Artech House, 2002.

[2] Shannon, C. E., "A Mathematical Theory of Communication," reprinted with corrections from *The Bell System Technical Journal,* Vol. 27, July, October 1948, pp. 379–423, 623–656.

[3] Anderson, B. J., and R. Johannesson, *Understanding Information Transmission,* IEEE Press, Wiley-Interscience, John Wiley and Sons, Inc., 2005.

[4] Hewlett Packard, *Digital Modulation in Communications Systems—An Introduction,* Application Note 1298, 1997.

[5] Sklar, B., *Digital Communications, Fundamentals and Applications,* Second Edition, Upper Saddle River, New Jersey: Prentice Hall, 2001.

[6] https://technet.microsoft.com/en-us/library/bb742566.aspx (accessed March 6, 2015).

[7] Larus Corporation, *Transmission Engineering Tutorials,* 1996.

[8] Flanagan W. A., *The Guide to T1 Networking,* Fourth Edition, Telecom Library Inc., 1990.

[9] ITU-T Recommendation G.702, *Digital Hierarchy Bit Rates,*1988.

[10] Horak, R., *Telecommunications and Data Communications Handbook*, Wiley-Interscience, John Wiley and Sons, Inc., 2007.

[11] ITU-T G.803, "Architecture of transport networks based on the synchronous digital hierarchy (SDH)," 03/2000.

[12] Goralski, W., *SONET,* Second Ed., New York: McGraw-Hill, 2000.

[13] Zhou, D., "Survivability in Optical Networks," *IEEE Network,* Nov./Dec. 2000.

[14] Lewin, B., "SONET Equipment Availability Requirements," In Global Telecommunications Conference and Exhibition, *Communications Technology for the 1990s and Beyond* (GLOBECOM), 1989. *IEEE,* pp. 1516–1521.

[15] Malis, A. G., "Reconstructing Transmission Networks Using ATM and DWDM," *IEEE Communications*, Vol. 37, No. 6, June 1999.

[16] Bates, J., *Optimizing Voice in ATM/IP Mobile Networks,* New York: McGraw-Hill, 2002.

[17] http://metroethernetforum.org/, accessed July 13, 2009.

[18] CISCO/Intel, *IEEE 802.3az - Energy Efficient Ethernet: Build Greener Networks,* White Paper, 2011.

[19] ITU-T E.800, *"Definitions of Terms Related to Quality of Service,"* 2008.

[20] Larus Corporation, *Digital Network Timing and Synchronization,* 1997.

[21] Achanta, S., and J. Hofer, "Considerations for Developing Resilient, High-Accuracy Time Synchronization Systems," *UTC Journal,* Fourth Quarter 2014.

[22] Federal Energy Regulatory Commission, Washington, D.C., USA. RM13-5-000, Version 5 Critical Infrastructure Protection Reliability Standards, April 2013. http://www.ferc.gov/whats-new/comm-meet/2013/041813/E-7.pd (accessed March 9, 2015).

[23] Department of Energy, *Communications Requirements of Smart Grid Technologies,* October 5, 2010.

[24] Ekanayake, J. et al., *Smart Grid Technology and Applications,* West Sussex, UK: John Wiley & Sons, 2012.

[25] MODBUS Organization, *MODBUS Application Protocol Specification*, V1.1b3, 2012.

[26] http://en.hartcomm.org/ (accessed April 06, 2015).

[27] IEEE Std 1615-2007, "IEEE Recommended Practice for Network Communication in Electric Power Substations," 2007.

CHAPTER 3

Introduction to Fiber-Optic System Design

3.1 Metric Units of Measurement

Readers of this book will notice the use of different units of measurement throughout the text; the main reason is the fact that the world today is a mixture of different units of measurement. Consider that formulas developed in North America use different units of measurement than those developed by the ITU in Switzerland.

The metric system is an international standard used by nearly all the countries of the world, while there are very few countries in the world still using the English system. The advantage of the metric system is the fact that it is based on powers of 10 so it is much easier to convert units—often just by moving the decimal point. Using the metric system, once you remember the meaning of the prefixes, you can easily convert mass, volume, and distance measurements—no need to memorize further conversion factors except the specific power of 10.

Historically, units of measurements pertaining to fiber-optic systems and components have used the metric system. This requires us to pay close attention to which units are used where and to make sure that the units are consistent in our calculations. It is impossible to overemphasize how important it is for engineers to be able to navigate between different systems of units of measurement (see Appendix B).[1]

3.2 Fiber-Optic Communication Theory

3.2.1 Optical Fibers

Optical fibers can be used to transmit light and thus information over long distances—they are widely used for telephony, but also for Internet traffic, long high-speed LANs, cable TV (CATV), power utilities, and increasingly also for shorter distances within buildings, control rooms, and even between pieces of equipment in

1. In September 1999, the United States lost the Mars Climate Orbiter as it approached Mars. The loss of the $125 million spacecraft was due to scientists confusing English units and metric units.

the same room. In most cases, silica fibers are used, except for very short distances, where plastic optical fibers can be advantageous.

Compared with systems based on electrical cables, the approach of optical fiber communications has advantages. The capacity of fibers for data transmission is huge—a single silica fiber can carry hundreds of thousands of telephone channels, utilizing only a small part of the theoretical capacity.[2] Due to the huge transmission rate achievable, the cost per transported bit can be extremely low.

The losses for light propagating in fibers are very small so that many tens of miles can be bridged without amplifying the signals. A large number of channels can be regenerated in a single fiber amplifier, if required for very large transmission distances. In addition, in comparison with electrical cables, fiber-optic cables are very lightweight.

Fiber-optic cables are immune to problems that arise with electrical cables, such as ground loops or EMI. Such issues are important for data links in industrial environments, including substations and other sites within power utility industry.

An optical fiber is a glass or plastic[3] fiber that carries light along its length. Optical fiber consists of a core, cladding, and a protective outer coating, which guides light along the core by total internal reflection. (See Figure 3.1.)

One of the disadvantages of optical fibers is the total cost of the equipment necessary to convert electrical signals to optical signals (e.g., optical transmitters and receivers). However, calculating cost per megabyte of data (a common practice in telecommunications), due to the very high amount of data that can be transmitted, it is by far the cheapest communication technology.

Joining sections of optical fiber is more complex than joining electrical wire or cable; the ends of the fibers must be carefully cleaved, and then spliced together, either mechanically or by fusing them with heat (fusion splicing), requiring specialized training and equipment.

Fibers are frequently classified by the number of modes they transmit. *Multimode fiber* allows the transmission of many modes; it is usually low in cost, but it is used for short distances and low data rates only. A *single-mode fiber* is efficient only for a single mode and is frequently used in communications that require long distances and high data rates.

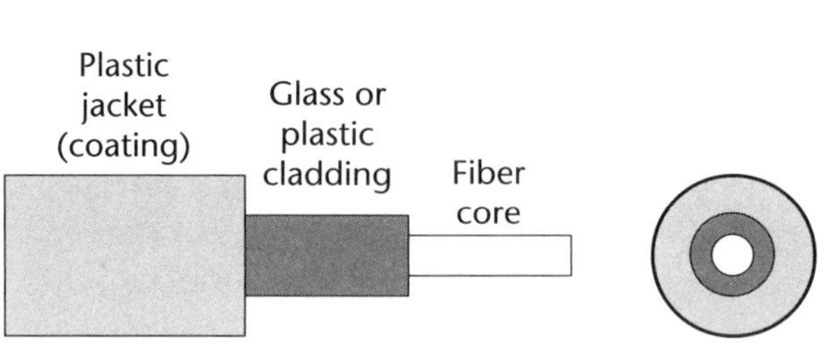

Figure 3.1 Optical fiber.

2. Over the last 30 years, the progress concerning transmission capacities of fiber links has been significantly faster than e.g. the progress in the speed or storage capacity of computers. Fiber optic product characteristics are in a constant state of change and it is wise to investigate before finalizing design and system specifications.

3. Plastic fibers are used only for short distances (for example, small LANs) due to higher attenuation (but easy to install), they better withstand stress, less expensive, and 60% less weight.

3.2.1.1 Multimode Optical Fibers

For short distances of a few hundred meters or less (e.g., within storage area networks), it is often more convenient to utilize multimode fibers, as these are cheaper to install (for example, due to their large core areas, they are easier to splice), and it is easy to successfully inject (in fiber jargon, this is called launch) pulses into them. Depending on the transmitter technology and fiber length, they achieve data rates between a few hundred megabits per second and ≈ 10 Gbps.

Multimode fibers have a fiber core diameter of either 50 or 62.5 microns. Multimode fiber introduces multimode distortion, which often limits the bandwidth and length of the link to few hundred meters. Multimode fibers were used more often in the past, but today they are almost completely being replaced with single-mode fibers; exceptions are systems inside the electrical substation and/or short distances of around a mile or less, where multimode fiber is still used.

3.2.1.2 Single-Mode Optical Fibers

Single-mode fibers are typically used for longer distances of a few kilometers or more. They have a fiber core diameter of 9–10 microns.[4] Currently used commercial telecom systems typically transmit 10 Gbps or 40 Gbps per data channel over distances of 10 kilometers or more. The newest available systems (as of 2014) reach 100 Gbps, although standards are already being developed for 400-Gbps systems.

The required total capacity is usually obtained by transmitting many channels with slightly different wavelengths through fibers; this is called *wavelength division multiplexing* (WDM). Total data rates can be several terabits per second, sufficient for transmitting many millions of telephone channels simultaneously. Even this capacity does not reach by far the physical limit of an optical fiber. In addition, note that a fiber-optic cable often contains multiple fibers.

Single-mode fiber [1] allows only one mode of light to propagate and thus eliminates the main limitation to bandwidth, modal dispersion. However, the small core of a single-mode fiber makes coupling light into the fiber more difficult so that lasers must be used. The main limitation to the bandwidth of the single-mode fiber is *chromatic dispersion*.[5]

3.2.2 Performance of Fiber-Optic Cables

3.2.2.1 Light Propagation Characteristics

Two main categories of optical fiber used in fiber-optic communications are multimode optical fiber and single-mode optical fiber. They have different characteristics and applications summarized as follows:

- 850 nm for shorter distances on multimode fiber;
- 1,300 nm for longer distances on multimode fiber;

4. For comparison purposes, the typical human hair has a diameter of 70 microns.
5. Chromatic dispersion is the variation of the index of refraction of the fiber with wavelength and can be a critical factor in DWDM (Dense Wavelength Division Multiplex) systems performance. The consequence of the chromatic dispersion is a broadening of the transmitted pulses.

- 1,310–1,320 nm for singlemode fiber;
- 1,500 nm (and above) is used for longer distances over singlemode fiber.

Table 3.1 shows different light propagation bands in which single-mode fibers are used. Each band has slightly different light propagation characteristics.

3.2.2.2 Bandwidth and Attenuation

There are two optical fiber factors that affect the transmission performance of a fiber-optic system: bandwidth and attenuation. Bandwidth is the measure of the data carrying capacity of the fiber, and the greater the bandwidth, the greater the information capacity. Bandwidth is expressed in a frequency-distance form (megahertz per kilometer, used only for multimode fibers) or as a maximum data rate expressed in gigabits per second for single-mode fibers. Bandwidth can be limited by fiber-optic equipment or by the dispersion of the optical fiber.

Attenuation describes the total loss (expressed in decibels) of an optical fiber system. As the light signal traverses the fiber, it will lose optical power (expressed in decibels per milliwatt) due to absorption, scattering, and other losses. At some point, the power level may become too weak for the receiver (reaching receiver threshold expressed in decibels per milliwatt) to distinguish between the optical signal and the background noise.

The *attenuation coefficient* (decibels per kilometer) depends on the wavelength of transmitted light, and for a uniform fiber it is independent of the length of the fiber.[6] It is specified by the manufacturer and provided for a specific fiber-optic cable.

3.2.2.3 Dispersion of Optical Fibers

As the light signal traverses the fiber, the light pulses will spread or broaden and limit the information-carrying capacity at very high bit rates or for transmission over very long distances. The overall effect of dispersion on the performance of a fiber optic system is known as intersymbol interference, which occurs when the pulse

Table 3.1 Light Propagation Bands

Band	*Wavelength Range*	*Description*
O band	1,260–1,360 nm	Original
E band	1,360–1,460 nm	Extended
S band	1,460–1,530 nm	Short wavelengths
C band	1,530–1,565 nm	Conventional ("erbium window")
L band	1,565–1,625 nm	Long wavelengths
U band	1,625–1,675 nm	Ultra-long wavelengths

6. In older fibers (e.g. ITU-T G.652.B) a strong peak of attenuation was present around 1383 nm due to the presence of the residual water vapor in silica (OH ion). In modern low water peak fibers (e.g. ITU-T G.652.D) the concentration of the OH ion is drastically reduced and the attenuation peak around 1383 nm has practically disappeared. The low water peak fibers allow the possibility of using a larger wavelength range.

spreading caused by dispersion causes the output pulses of a system to overlap, rendering them undetectable. There are two types of dispersion: chromatic dispersion and modal dispersion. Chromatic dispersion consists of material dispersion and waveguide dispersion [2].

Waveguide dispersion occurs because the mode propagation constant is a function of the size of the fiber's core relative to the wavelength of operation. Waveguide dispersion also occurs because light propagates differently in the core than in the cladding.

Material dispersion occurs because the spreading of a light pulse is dependent on the wavelengths' interaction with the refractive index of the fiber core. Material dispersion is a function of the source spectral width, which specifies the range of wavelengths that can propagate in the fiber. Material dispersion is lower at longer wavelengths.

Intermodal, or modal, dispersion occurs only in multimode fibers and it contributes to pulse broadening.

Polarization mode dispersion (PMD) is a special case of modal dispersion (usually associated with single-mode fibers) where two different polarizations of light in a waveguide travel at different speeds due to random imperfections and asymmetries, causing random spreading of optical pulses. PMD makes it impossible to transmit data reliably at high speeds, and it is especially difficult to deal with. Its effects prevent many existing optical systems from using high-bandwidth transmission equipment meeting 10 Gbps (OC–192) or higher specifications.

PMD affects the transmission quality by spreading signal pulses (and distortions in analog systems) in digital systems and therefore, raising the BER of the system. For long optical fibers, the figure of merit usually specified is *PMD coefficient.*

The PMD for an installed fiber-optic cable is dominated by the inherent PMD of the bare fiber; however, the level of relaxation provided by the cabling and installation techniques also affects PMD [3]. In the design of a reliable, long-haul fiber-optic network, the relationship between the maximum achievable link length and bit rate is influenced by PMD and must therefore be considered. The practical expression[7] is:

$$L = \frac{10^4}{B^2 \cdot PMD_{coeff}^2} \tag{3.1}$$

where L is maximum transmission length in kilometers, B is a bit rate in gigabits per second, and PMD coefficient is expressed in $\frac{ps}{\sqrt{km}}$, where ps means picoseconds.

3.3 Fiber-Optic System Components

3.3.1 Basic Building Blocks of the Fiber-Optic System

The simplest type of fiber-optic communications system is a fiber-optic link providing a point-to-point connection with a single data channel. Three primary

7. Based on the 2006 JDSU white paper "Testing Polarization Mode Dispersion (PMD) in the Field."

components in a fiber-optic link are the optical transmitter, optical receiver and fiber-optic cable.

In the transmitter, the input electrical signal modulates the light output from the semiconductor laser diode, which is then focused into the fiber-optic cable. This fiber cable carries the modulated optical signal to the receiver, which then reconverts the optical signal back into the original electrical (analog or digital) signal. Figure 3.2 simply illustrates the fiber-optic communications system.

The *optical transmitter* produces and encodes the light signals; it is based mostly on semiconductor lasers, fiber lasers, and optical modulators. The *optical fiber* conducts the light signals over a distance; fibers have optimized properties concerning losses, guiding properties, dispersion, and nonlinearities. The *optical receiver* receives and decodes the light signals, mostly based on photodiodes (often avalanche photodiodes). An *optical regenerator* may be necessary to boost the light signal at long distances.

Some other important components of the fiber-optic system are dispersion-compensating modules; semiconductor and fiber amplifiers (mostly erbium-doped fiber amplifiers, sometimes Raman amplifiers) for maintaining sufficient signal powers over long lengths of fibers, or as preamplifiers before signal detection; optical filters (e.g., based on fiber Bragg gratings) and couplers; optical switches and multiplexers; electrically controlled optical switches; various kinds of electronics (e.g. for signal processing and monitoring); and computers and software to control the system operation.

This text only briefly describes optical transmitters and receivers.

3.3.1.1 Optical Transmitter

The optical source must be able to generate enough optical power so that the desired BER can be met over the optical path. There should be high efficiency in coupling the light generated by the optical source into the fiber-optic cable, and the optical source should have sufficient linearity to prevent the generation of harmonics and intermodulation distortion.

In addition, the optical transmitter must modulate this light to represent the binary data that it receives from the source. The optical source must be easily modulated with an electrical signal and must be capable of high-speed modulation. The transmitter is typically pulsed at the incoming frequency and performs electrical-to-optical (EO) conversion. *Light-emitting diodes* (LEDs) or *vertical cavity surface emitting lasers* (VCSELs) are used to drive multimode fiber systems, whereas laser diodes (LDs) are used to drive singlemode fiber systems.

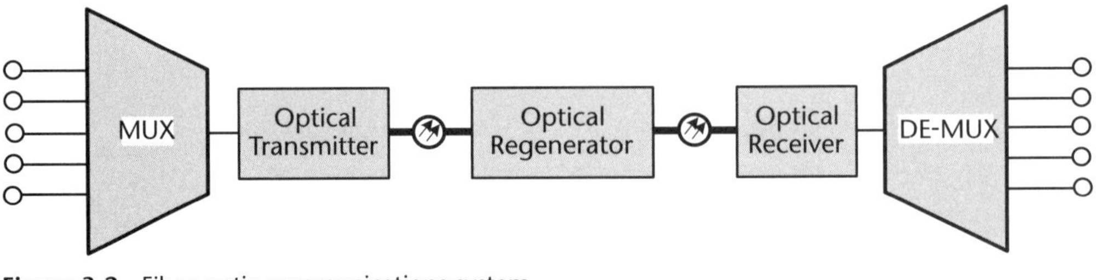

Figure 3.2 Fiber-optic communications system.

A key difference between the optical output of an LED and a LD is the wavelength spread over which the optical power is distributed. The *spectral width*, (σ), is the 3-dB optical power width (measured in nanometers or microns). The spectral width impacts the effective transmitted signal bandwidth. An LD always has a smaller spectral width than an LED.

There are several different schemes for carrying out the modulation function but within the context of a premise fiber-optic data link, the only one really used is *intensity modulation* (IM). The amount of power that the LED or LD radiates (sometimes referred to as the radiance) is proportional to the input current. If the input current is the waveform representing the binary information stream, the resulting optical signal looks like bursts of optical signal when input waveform represents a 1 and the absence of optical signal when input waveform represents a 0. This is also known as *direct modulation* of the LED or LD.

3.3.1.2 Optical Receiver

The receiver serves two functions—it must sense or detect the light coupled out of the fiber-optic cable and convert the light into an electrical signal, and it must demodulate the binary data that it represents. The receiver performs the optical-to-electrical transducer function.

The light detection is carried out by a photodiode, which senses light and converts it into an electrical current. However, the optical signal from the fiber-optic cable and the resulting electrical current will have very small amplitude, and it will have to be amplified by one or more amplification stages, including filters and equalizers to shape and improve the information-carrying electrical signal. The receiver can incorporate a number of other functions, such as clock recovery for synchronous signaling, decoding circuitry, and error detection and recovery.

There are two types of photodiode structures: *positive intrinsic negative* (PIN) and the *avalanche photodiode* (APD). In most premise applications, the PIN is the preferred element in the receiver. This is mainly due to fact that it can be operated from a standard power supply, typically between 5 and 15V.

APD devices have much better sensitivity and twice the bandwidth; they are usually found in long-haul communication links and in metro-regional networks. However, they cannot be used on a 5-V printed circuit board. They also require a stable power supply, which increases their cost.

3.3.1.3 Regenerators and Amplifiers

In a traditional metallic analog environment, transmitted signal weakens over distance, so amplifiers are placed in the circuit to raise the power level of the signal. Unfortunately, in addition to amplifying the signal, amplifiers also amplify whatever cumulative noise has been picked up by the signal during its trip across the network. At some distance, it becomes difficult for a receiver to discriminate between the actual signal and the noise embedded in the signal and very complex circuitry is required to discriminate between the two.

In digital systems, *regenerators* are used not only to amplify the signal, but also to remove any noise and signal distortion that has been picked up along the way. Regenerators, as the name implies, have the job of regenerating the clock and amplitude relationships of the incoming data signals, which have been attenuated and distorted by dispersion. They derive their clock signals from the incoming data stream.

In a traditional installation, the optical signal is received by a receiver circuit, converted to its electrical analog, regenerated, converted back to an optical signal, and transmitted forward over the next fiber segment. This optical-to-electrical-to-optical conversion process is costly, complex, and time-consuming but necessary.

The newer approach is to use *optical amplifiers*, which receive the optical signal, amplify it as an optical signal, and then retransmit it as an optical signal with no electrical conversion required. Optical amplifiers[8] are actually rather simple devices that, as a result, tend to be extremely reliable; since, however, they also amplify the noise, at some point, signal regeneration may still be required.

In *Erbium-doped fiber amplifiers* (EDFAs), the coil of doped fiber lies at the heart of the optical amplifier's functionality (Figure 3.3). Doping is simply the process of embedding some kind of functional impurity in the silica matrix of the fiber when it is manufactured. In optical amplifiers, this impurity is often erbium.

The pump laser generates a light signal at a particular frequency, often 980 nm, in the opposite direction than the actual transmitted signal flows. As it turns out, erbium becomes atomically excited when it is struck by light at that wavelength. When an atom is excited by pumped energy, it jumps to a higher energy level and then falls back down, giving off a photon within the 1,550-nm region. So, when the weak transmitted signal reaches the coil of erbium-doped fiber, the erbium atoms excited by the energy from the pump laser bleed power into the weak signal at precisely the right wavelength, causing an amplification of the transmitted signal. The optical isolators serve to prevent errant light from backscattering into the system. The maximum number of erbium-doped amplifers in a fiber chain is about four to six.

In another option to amplify optical signals, *Raman amplification*, the signal beam (data) travels down the fiber alongside a rather powerful pump beam, which excites atoms in the silica matrix that in turn emit photons that amplify the signal. The advantage of Raman amplification is that it requires no special doping; erbium is not necessary. Instead, the silica itself gives off the necessary amplification. In this case, the fiber itself becomes the amplifier.

Raman amplifiers require a significantly high-power pump beam, and they will operate within a relatively wide range of wavelengths, including 1,310- and

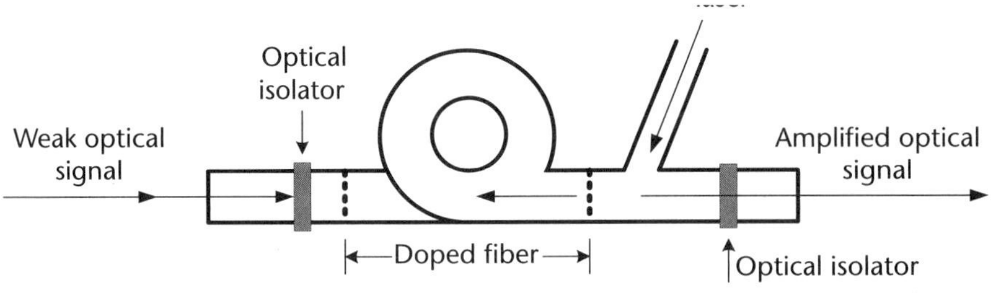

Figure 3.3 Erbium-doped fiber amplifier.

8. The technologically most important rare-earth-doped fibers are erbium-doped fibers for erbium-doped fiber amplifiers and ytterbium-doped fibers for high-power fiber lasers and amplifiers.

1,550-nm, currently the two most popular and effective optical transmission windows.

3.3.2 SONET and SDH

SONET and SDH are standardized protocols that transfer multiple digital bit streams synchronously over optical fiber. At the interface, SONET converts digital signals from electrical form to optical form and back to electrical form at the destination. It supports a transmission range from 51.840 Mbps to multiples of DS3s such as OC-248 (9.9 Gbps) and the multiplexing of lower capacity channels, down to the 64 kbps (so-called DS0) channel [4].

The SONET standard includes four functional layers: the *photonic*, the *section*, the *line*, and the *path layer*, described as follows. [They correspond to both the physical and the data link layers (Figure 3.4).]

- The photonic layer is the physical layer of SONET that deals with cable, signal, and component specifications such as optical fiber's minimum power requirements and the dispersion characteristics of transmitting lasers. Signals are converted between electrical and optical forms in this layer.
- In the section layer basic frames are created and scrambled whenever appropriate. It converts electrical signals to photonic ones and monitors for errors.
- The line layer is responsible for getting frames from one end of a line to the other. The synchronization[9] and multiplexing of data onto the SONET frames are done in this layer. This means that the timing adjustments and adding or dropping of the traffic are made at this level. It also does protection, maintenance functions, and switching.

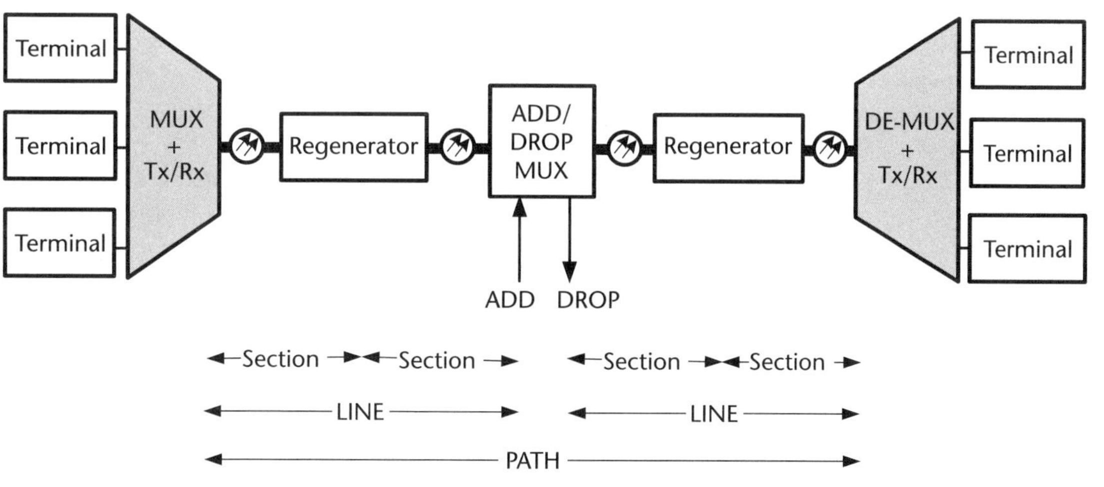

Figure 3.4 SONET network layers.

9. SONET networks require that the timing at all nodes be synchronized to a Stratum 1 clock or Primary Reference Source (PRS). For small, or non-critical, networks one PRS is typically used in the interest of cost. For larger, or critical, networks, a PRS is also deployed at one SONET core node, but a secondary (standby) PRS is also connected to one of the other nodes in the SONET network. Secondary PRS is not used until the primary PRS loses Stratum 1 timing for any reason, after which the secondary PRS clock takes over.

- The path layer completes the transmission (e.g., getting from one end to another with overall path, this layer provides for transport of data at the appropriate signaling speed). A single-stage multiplexer/demultiplexer can multiplex various inputs[10] into an OC-N signal. At an add/drop site, only those signals that need to be accessed are dropped or inserted. The remaining traffic continues through the network element without requiring special pass-through units or other signal processing. In other words, the add/drop multiplexer provides interfaces between the different network signals and SONET signals.

The North American SONET system is based on multiples of a fundamental rate of 51.840 Mbps, called STS-1 (for synchronous transmission signal, level 1).[11] For example, STS-1 = 51.840 Mbps, STS-3 = 155.520 Mbps, STS-9 = 466.560 Mbps, STS-12 = 622.080 Mbps, and STS-48 = 2,488.320 Mbps.

Optical carrier (OC-*n*) levels describe a range of optical digital signals that can be carried on the SONET network. The general rule for calculating the speed of optical carrier lines is that when a specification is given as OC-*n*, the speed will equal $n \times 51.840$ Mbps. The facility designators indicate the facility type, which is usually fiber-optic cable (e.g., OC-1 is an optical carrier supporting an STS-1 signal, whereas OC-3 supports a STS-3 signal and so forth).

A portion of the bits transferred over the line is designated as *overhead*. The overhead carries information that provides operations, administration, maintenance, and provisioning (OAM&P) capabilities such as framing, multiplexing, status, trace, and performance monitoring. The payload rate (line rate – overhead rate = payload rate) is the bandwidth available for transferring user data.

SONET is an ANSI standard; it can carry as payloads the North American PDH hierarchy of bit rates—1.5/6/45 Mbps, plus 2 Mbps. SDH embraces most of SONET and is an international standard, but it is often regarded as European standard because its suppliers, with few exceptions, carry only the ETSI-defined European PDH bit rates of 2/34/140 Mbps (8 Mbps is omitted from SDH).[12]

3.3.3 Dense Wavelength Division Mutliplexing (DWDM)

DWDM[13] is a fiber-optic transmission technique that employs light wavelengths to transmit data parallel-by-bit or serial-by-character and can increase the bandwidth of the existing fiber-optic facilities presently up to 64 times.

In order to squeeze more bandwidth out of their fiber networks, users are deploying DWDM to build backbones that might have dozens of channels

10. Although network elements (NEs) are compatible at the OC-N level, they may differ in features from vendor to vendor. SONET does not restrict manufacturers to providing a single type of product, nor require them to provide all types. For example, one vendor might offer add/drop multiplexer with access at DS1 only, whereas another might offer simultaneous access at DS1 and DS3 rates.
11. The international SDH system is based on a fundamental rate of 155.520 Mbps, three times that of the SONET system. This fundamental signal is called STM-1 (synchronous transmission module, level 1).
12. European Telecommunications Standards Institute (ETSI) defined SDH for Europe. It is now used everywhere outside North America and Japan. It was synchronous and based on byte interleaving and not on bit interleaving as was the legacy PDH.
13. In the late 1990s there was an explosion of development activity aimed at deriving benefit of the concept of DWDM.

(wavelengths) riding on a single strand of fiber, with each channel operating at multigigabit speeds. (See Figure 3.5.) Terminals 1 and 2 may be different applications at the same site, or the same traffic intended for different sites.

DWDM increases the capacity of embedded fiber by first assigning incoming optical signals to a specific wavelength within a designated frequency band and then multiplexing the resulting signals out onto one fiber. Because incoming signals are never terminated in the optical layer, the interface can be bit-rate and format-independent, allowing the user to integrate the DWDM technology easily with existing equipment in the network while gaining access to the untapped capacity in the existing fiber.

The most common form of DWDM uses a fiber pair—one for transmission and one for reception. Systems do exist in which a single fiber is used for bidirectional traffic, but these configurations must sacrifice some fiber capacity by setting aside a guard band to prevent channel mixing.

Intermediate optical regenerators in DWDM systems may allow for the dropping and adding of certain wavelength channels but this is done infrequently, because adding or dropping wavelengths requires manually inserting or replacing wavelength-selective cards. This is costly and, in some systems, requires that all active traffic be removed from the DWDM system, because inserting or removing the wavelength-specific cards interrupts the multiwavelength optical signal.

With a new type of multiplexer—*reconfigurable optical add-drop multiplexer* (ROADM)—network operators can remotely reconfigure the multiplexer by sending soft commands. The architecture of the ROADM is such that dropping or adding wavelengths does not interrupt the traffic on other wavelengths.

3.3.4 Course Wavelength Division Multiplexing (CWDM)

CWDM is a method of combining multiple signals on laser beams at various wavelengths for transmission along fiber-optic cables, such that the number of channels is fewer than in DWDM but more than in standard WDM.

CWDM systems have channels at wavelengths spaced 20 nm apart, compared with 0.4-nm spacing for DWDM. This allows the use of low-cost, uncooled lasers for CWDM. In a typical CWDM system, laser emissions occur on eight channels at eight defined wavelengths—1,610; 1,590; 1,570; 1,550; 1,530; 1,510; 1,490; and

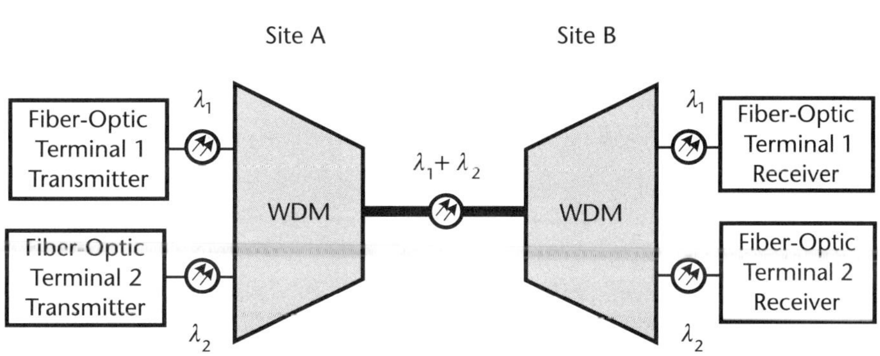

Figure 3.5 Fiber-optic capacity expansion.

1,470 nm—although up to 18 different channels are allowed, with wavelengths ranging down to 1,270 nm.

The energy from the lasers in a CWDM system is spread out over a larger range of wavelengths than is the energy from the lasers in a DWDM system. The tolerance (extent of wavelength imprecision or variability) in a CWDM laser is up to ±3 nm, whereas in a DWDM laser the tolerance is much tighter. Because of the use of lasers with lower precision, a CWDM system is less expensive and consumes less power than a DWDM system. However, the maximum realizable distance between nodes is smaller with CWDM.

3.4 Fiber-Optic Systems in Electric Utility Space

Fiber-optics communication offers several advantages over metallic systems. One of them is that any form of outside electronic, magnetic, or radio frequency interference does not affect the transmitted signals. Therefore, optical cables are completely immune to lightning or high-voltage interference. Furthermore, optical fibers will emit no radiation, which ideally suits them for today's tougher standards for interference, also known as electromagnetic compatibility (EMC).

Because optical signals do not require grounding connections, the transmitter and receiver are electrically isolated and free from ground loop problems. With no chance of terminal-to-terminal ground potential shifts, plus safety from sparking and shock, fiber-optic systems are the choice for many processing applications where safe operations in hazardous or flammable environments is a requirement [5].

In most applications, optical fiber must be protected from the environment using a variety of different cabling types based on the type of environment in which the fiber will be used. Cabling provides the fiber with protection from the elements, added tensile strength for pulling, rigidity for bending, and durability. In general, fiber optic cable can be separated into two types—indoor and outdoor.

Optical fibers are particularly suitable for use as transmission media by means of the aerial power lines in high-voltage networks, offering low attenuation (long distance between repeaters), large bandwidth (high transmission capacity), and immunity to electromagnetic influences and cross-talk.

Optical cables are installed in various environments (e.g., aerial, buried, duct, tunnel, and underwater) and are therefore exposed to different environmental conditions. The range of environmental conditions must be considered with great care in order to determine the cable construction that will continuously maintain the desired characteristics. The external factors relating to the various environmental conditions can be divided into two categories: natural external factors (e.g., temperature, wind, water, and earthquakes and man-made factors (e.g., smoke, air pollution, and fire).

For the utility communications system[14], fiber-optic cable is commonly installed on transmission or distribution lines using OPGW, OPPC, and ADSS cable

14. Aside from using optical fibers for communications, there are a number of other applications in gas, oil, and electrical utilities; for example, sensing applications where physical measurement of temperature or pressure and tensile forces, can affect glass fibers and locally change the characteristics of light transmission

installed on transmission line towers, or fiber-optic cable supported by a metallic messenger (lashed or figure eight–style cables). Inside the substation fiber-optic cables are usually put in the underground duct.

3.4.1 OPGW

OPGW is a transmission line overhead ground wire[15] with optical fibers. It is primarily used by the electric utility industry and represents a dual-functioning cable, designed to replace traditional ground wires on overhead transmission lines with the addition of optical fibers, which can be used for communications purposes (Figure 3.6). However, if the OPGW cable fails it may not be repaired quickly. Therefore, an alternative routing for the optical circuits needs to be considered.

As OPGW functions as both electrical and optical transmission media, its design and construction has to satisfy both electrical and optical specifications. OPGW cables consist of a core containing optical fibers and outside armor generally composed of one or more layers of aluminum wire, Aldrey metallic wire, steel wire or aluminum-coated steel wire.

The two major electrical requirements of the cable are low DC resistance and fault current withstand capability. OPGW must also be capable of handling electrical faults on the transmission line by providing a path to ground without damaging the sensitive optical fibers inside the cable. Particular attention should be given on fiber strain as the cable is under tension throughout its life. OPGW[16] must be capable of withstanding the mechanical stresses applied to overhead cables by environmental factors such as wind and ice. The performance, test requirements, procedures, and acceptance criteria for an OPGW are covered in [6].

Optical signals can become attenuated to the point that they must be regenerated typically every 80 km (50 miles). There are solutions by some fiber-optic equipment manufacturers that promise much longer runs without a need for optical signal regeneration, but these are nonstandard, and in some cases, proprietary, solutions that may not be suitable for all types of utility applications.

Utilities typically install many more fibers than they need for its internal communications in order to allow for future expansion and to lease or sell to telecommunications companies. Rental fees for these *dark fibers*[17] can provide a valuable additional source of revenue for the electric utility. However, when *rights-of-way* (ROW) for a transmission line have been expropriated from landowners, occasionally utilities have been restricted from such leasing agreements on the basis that the original ROW was only granted for electric power transmission.

---

in the fiber. Distributed temperature sensing systems (DTS) are optoelectronic devices which measure temperatures by means of optical fibers functioning as linear sensors. DTS can be used to monitor the power cable's condition, identify hotspots and bottlenecks, and to minimize risks and costs for maintenance and in emergency situations.

15. Ground wire is also known as shield wire, static wire, earth wire, or skywire.

16. OPGW was invented in 1970s but developed for practical use in early 1980s, after realization that the electric transmission lines provided an ideal right-of-way for a communications path. Fibers for OPGW are singlemode type.

17. Dark fibers refer to unused fibers; optical fiber conveys information in the form of light pulses so the "dark" means no light pulses are being sent.

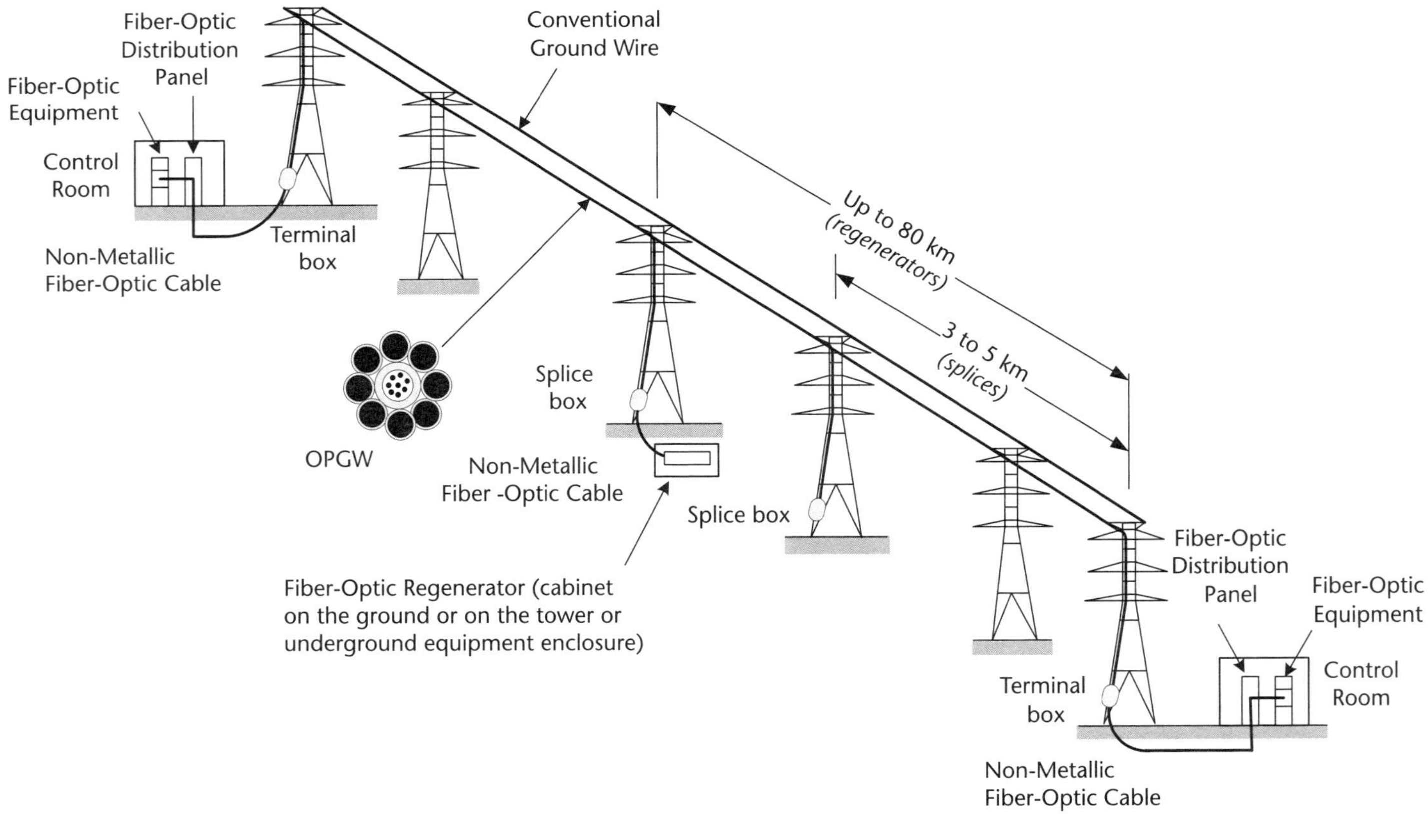

Figure 3.6 Optical ground wire.

OPGW installation is basically different from that of other types of cable. Although OPGW installation is typically not the responsibility of the communications engineers, some basic understanding of this process is essential. The following factors should be considered in determining the type of cable, maximum tension, and the installation plan:

- Maximum short-circuit current through the cable.
- Disconnection time of a short-circuit to Earth.
- Sag of the phase conductors.
- Spans and their positions in relation to poles.
- Maximum wind speed.
- Maximum ice load.
- Other aspects, such as risk of atmospheric discharge, fire, discharge of birdshot, saline, fog, and aggressive chemical agents in the atmosphere.
- OPGW—cable pulling precautions:
    - Always keep reels in a vertical position with the ends secured; inspect following transportation to ensure that no damage has occurred.
    - Always turn in the direction indicated on the reel.
    - During the cable pulling, sufficient mechanical tension should be maintained to prevent the cable from touching the ground or any other obstacle. Control the speed and maximum tension of stringing so that the recommended values are not exceeded.
    - Proper accessories and tools are important as the optical fibers are very sensitive and must not be compressed or bent too sharply and the recommended minimum bending radius should be followed.
- When splicing takes place on the ground, the splice box is subsequently raised to the appropriate height on the pole (10-ft minimum) and the cable being wound and fastened.

Quite often, the electric and mechanical properties of the OPGW are designed and customized to meet the user's specific requirements. Communications engineers will be responsible for the very important transmission characteristics that affect the use of the fibers—the attenuation and the dispersion of light.[18] A low attenuation of light is needed for long repeater spacing, and a low dispersion is needed for high-speed data transmission [7].

Testing fiber-optic cable four times—each reel after manufacturing, each reel after delivery, each section after installation and splicing, and final end-to-end testing after all sections are installed and spliced—is recommended. (See Chapter 7 for more details.)

18. Length of most fiber-optic cables is not the same as the length of the optical fibers inside; manufacturers put about one percent more fiber in the cable than the length of the cable itself, to allow for some expansion. This could cause some confusion during the testing and/or troubleshooting the problems on the longer fiber-optic cable sections.

3.4.2 Optical Phase Conductor (OPPC)

OPPC is not very common in North America, but it can be used as an alternative communications solution when there is no existing ground wire and OPGW is not a viable option. The basic construction is similar to conventional OPGW, only it is designed to simulate the mechanical and electrical characteristics of the phase wire it replaces.

Unlike OPGW, where the cable is not carrying continuous current, OPPC is energized along high-voltage power lines, and therefore it requires specially adapted splice boxes and termination units at both ends of the line to accommodate the live line conditions. In addition, OPPC can be used in addition of the existing OPGW to provide a secondary (or backup) communication path.

3.4.3 ADSS

ADSS cable is ideal for installation in distribution as well as transmission environments, even when live-line installations are required. It does not require support or messenger wire, so installation is achieved in a single pass, making ADSS an economical and simple means of building a fiber-optic network.

Aerial fiber-optic cable is exposed to many external forces; wind and ice loading can put stresses on cables, and temperature changes are more extreme than in any other installation environment. Furthermore, vehicular accidents, rodents, lightning, and even gunshots can damage aerial cables.

All-dielectric cables have no conducting materials within the cable, eliminating the need for grounding or surge suppression at the building entrance. These cables can be run parallel to electrical service cables without danger of induced voltages and in the vicinity of power lines.

All-dielectric outside plant cables are for aerial applications and duct applications where rodents are not a threat. Kevlar may be used for protection of all-dielectric cables if there is a threat of rodents.

More details about performance and testing of ADSS cables can be found in [8].

3.4.4 Metallic Aerial Self-Supporting (MASS) Cable

MASS cables are an alternative solution used for installing optical cable on medium- and high-voltage power lines when existing phase or ground wire replacement is not possible or economical. MASS cable is a compact, lightweight solution with no electrical function, designed to provide a telecommunications path without interfering with the existing power lines or infrastructure. Its small size helps minimize loading on towers and poles, yet it is completely self-supporting to meet sag and tension requirements, and typically installed in “under-build” applications beneath the live phases.

3.4.5 Buried Fiber-Optic Cables

Fiber-optic cable can also be buried in the ground; fiber-optic cable is usually (but not always) installed in an inner duct[19] that has been placed in a standard duct in advance of the fiber optic cable placing operation. Normally, fiber-optic cables designed for direct buried deployments have a galvanized steel wire for strength and a steel armor to protect them against crushing forces and rodents.[20]

Fiber-optic cables entering electrical utility premises have to be nonmetallic, so the cable protection is achieved by use of water-blocking, nonmetallic strength member, a special outer jacket, and a dielectric central strength member. *Fiber reinforced plastic* (FRP) strength member, which is dielectric in nature, is well-suited for cables surrounded by large electric fields like those found near electric power lines and transformers. A means of locating the cable, especially all-dielectric cable, should be provided.

Armored cables should be used when the cable will be direct buried in rocky soil or when rodent protection is required. PVC ducts with an outside diameter greater than 1.5 inches will also provide rodent resistance where armor cannot be used.

It is recommended that cables be buried as deep as possible, at least 30 inches underground, if possible, and located below the frost line to prevent damage due to frost heave.

3.4.6 Undersea Fiber-Optic Systems

Probably the most advanced optical communications networks are those used for undersea (submarine) communication.[21] Optical submarine cable systems with optical regenerators became mainstream after the fiber optics and the semiconductor lasers were put to commercial use in the 1980s. In the 1990s, transmission capacity has been increased dramatically by adopting both optical amplification and WDM transmission technologies [9].

In a land-based network the most significant cost is associated with digging a trench in which to bury fiber. Accordingly, when such a trench is created, it is wise to put a lot of fibers into it, since additional fibers do not cost much and will offer extra capacity for the future expansion. In the undersea situation things are quite different; amplifiers (or repeaters) are required at certain intervals, and these are costly to install and to maintain. As a result, in the undersea environment, there is a very large financial incentive to minimize the number of fibers in the cable. When the fiber-optic cable is laid in the sea, one or two spare fibers might be useful but a large number of additional fibers would be useless unless the (presently unneeded)

19. Groups of two to four 25-mm (1 inch) diameter innerducts are commonly routed through 100-mm (4 inch) diameter ducts.

20. Gophers are a real problem for fiber-optic cables in the U.S. There is actually a standardized test (conducted by a nature and wildlife organization) which involves placing a cable in a gopher enclosure for a fixed, specified length of time. Rodent-resistance of the cable is provided by coated steel tape armor. In other countries termites have been known to attack and eat the plastic sheathing.

21. The submarine cable system has a very long history; the installation of the world's first submarine cable was at Dover Strait in 1850.

amplifiers and other equipment to operate them in the future are installed at the same time.

Manufacturing submarine cables capable of transmitting electricity reliably for a long period under the harsh natural conditions at the bottom of the sea requires extremely advanced technology. In addition, undersea fiber-optic cables for utility applications are sometimes combined with power cables, which, on shorter runs, (i.e., no undersea regenerators required) can be a beneficial technique.

Even these undersea fiber-optic communications channels should have a redundant physical path, especially since a repair of the submarine fiber-optic cable could take days to complete.

3.4.7 Intrabuilding Fiber-Optic Cables

In general, the National Electric Code (NEC) identifies three different intrabuilding regions with regard to cable placement: plenums, risers, and general-purpose areas.

A *plenum area* is a compartment or chamber that forms part of the air distribution system and to which one or more air ducts are connected. A room with a primary function of air handling is also considered a plenum space. A *riser* is an opening or shaft through which cable may pass vertically from floor to floor in a building. *General purpose* areas are other indoor areas that are not plenum or risers. Cables are specifically listed for use in each of these environments.

The NEC does allow the use of a cable with a more stringent listing (more expensive as well) to be used in an application requiring a lesser listing but not the other way around. When a cable run will be exposed to both indoor and outdoor environments, there are several options to consider, each with its advantages and disadvantages. The first option is to run an outside plant cable for the entire run (may be too rigid for indoor runs). The second option is to use an indoor/outdoor cable. The third option is to transition-splice the outside plant cable to an inside plant cable.

An optical fiber connector terminates the end of an optical fiber and enables quicker connection and disconnection than splicing. The connectors mechanically couple and align the cores of fibers so light can pass through; good connectors lose very little light due to reflection or misalignment of the fibers. Aside from commonly found LC, SC, FC, and ST types, fiber-optic connector types include FDDI, ESCON, MU, D4, Biconic, SMA, MTP, and many others.

Since its inception in the 1960s, fiber-optic systems have utilized hundreds of different fiber-optic connectors. The following is a brief description of some of the most commonly used fiber-optic connectors today (Figure 3.7).

LC (licensed by Lucent Technologies) is a small-form factor connector that uses a 1.25-mm ferrule, half the size of the SC. Its good performance is highly favored

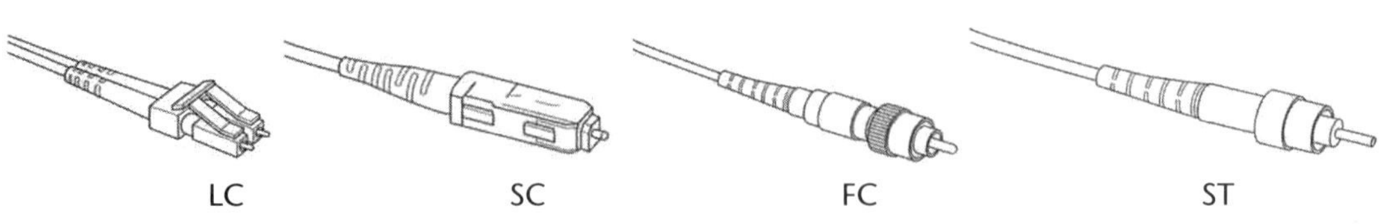

Figure 3.7 Fiber-optic connectors.

for singlemode fibers; it is a high-density optical connector of choice, and available in simplex or duplex versions.

SC is a new, third-generation connector with a 2.5-mm ferrule that is widely used for its excellent performance. It was the connector standardized in TIA-568-A, but it was not widely used at first because it was more expensive than a ST. It is a snap-in connector that latches with a simple push-pull motion. It is also available in a duplex configuration.

Face Contact (FC), originally designed by NTT (Nippon Telephone & Telegraph), was one of the most popular single-mode connectors for many years and commonly used with SONET and SDH equipment. It uses a 2.5-mm ferrule, but some of the early ones use ceramic inside stainless steel ferrules. It screws on firmly, but you must make sure you have the key aligned in the slot properly before tightening. It's mostly been replaced by SCs and LCs.

Straight tip (ST), an AT&T Trademark, is probably still the most popular connector for multimode networks, like most buildings and campuses and inside of the substations. It has a bayonet mount and a long cylindrical 2.5-mm ceramic (usually) or polymer ferrule to hold the fiber. Most ferrules are ceramic, but some are metal or plastic. Because STs are spring-loaded, it is important to make sure they are seated properly.

Table 3.2 shows typical losses for fiber-optic connectors.

Outside plant applications may require connectors to be located underground, or on outdoor walls or utility poles. In such settings, protective enclosures are often used, and their selection depends on the cable and connector type, the location, and environmental factors.

Fiber-optic *patch cords (jumpers)* are short fiber-optic cables (a few feet or a few tens of feet long) connecting an outside plant and master distribution frame with the indoor fiber-optic equipment. Patch cord cables are available for all building applications—plenum, riser, and general purpose.

Multimode and single-mode *patch cords* can be *simplex* or *duplex*. Simplex has one fiber, while duplex patch cord has two fibers joined with a thin web. Because simplex fiber-optic cable consists of only one fiber link, they are used for applications that only require one-way data transfer. Duplex multimode or single-mode fiber-optic cables are used for applications that require simultaneous, bidirectional data transfer. Workstations, fiber switches and servers, Ethernet switches, backbone ports, and similar hardware require duplex cable.

Pigtails (cables with an optical fiber connector on one end only) are used for terminating cables. A fiber optic pigtail is typically a one-or two-fiber cable that has been connectorized on one end with a fiber connector while the other end

Table 3.2 Fiber-optic Connector Losses

	Multimode Fiber		*Single-Mode Fiber*	
Connector Type	*Insertion Loss (decibels)*	*Return Loss (decibels)*	*Insertion Loss (decibels)*	*Return Loss (decibels)*
LC	0.10	> 30 dB	0.20	> 50 dB
SC	0.75	> 20 dB	0.75	> 26 dB
FC	0.50	> 20 dB	0.50	> 50 dB
ST	0.50	> 20 dB	0.50	> 50 dB

remains unterminated. This unterminated end is spliced to the cable that requires termination. After splicing, the splice point is protected by a splice tray and placed into terminating hardware.

3.4.8 Small Form-Factor Pluggable (SFP) Modules

The SFP module is a compact, hot-pluggable transceiver used for both telecommunications and data communications applications (Figure 3.8). It is a popular industry format jointly developed and supported by many network component vendors, and conforms to the Small Form-Factor Pluggable Multisource Agreement (MSA).

SFP transceivers are designed to support SONET, gigabit Ethernet, fiber channel, MPLS, and other communications equipment and standards.

3.4.9 Optical Power Budget

The key to fiber-optic link distance is optical power budget—the amount of light available to make a fiber optic connection (Figure 3.9). A simple calculation is used to determine how much fiber-optic light (in dBm),[22] is available [10].

The first step in calculating the optical power budget is determining how much light is available for the electronic devices themselves. Two measurements are needed from the manufacturer of the equipment—*minimum transmit power* and *minimum receiver sensitivity minimum transmit power*[23] represent the worst-case transmit power for a device guaranteed to provide at least that much power.

The second piece of information required is the *minimum receiver sensitivity*. Minimum receiver sensitivity is usually a negative number and represents the minimum amount of light required by the receiver to operate correctly. (Again, the actual minimum should be used, not an average of minimums.)

With minimum transmit power and minimum receive sensitivity data, it is possible now to calculate the *available optical power* as a "worst case scenario." In reality, the available power[24] will probably be higher than this calculated value:

$$\text{Available Optical Power} = \text{Minimum Transmit Power} - \text{Minimum Receiver Sensitivity} \quad (3.2)$$

Figure 3.8 SFP modules.

22. dBm tells us how many decibels over 1mW of electrical, or in this case optical power is available.
23. It is important to note that some vendors will list an "average minimum transmit power". Average minimum transmit power does not guarantee that a product will perform at that product's minimum transmit power.
24. When connecting devices from different vendors, or different models of products from the same vendor, the available power calculation needs to be computed in both directions with the smaller of the two used for the rest of the calculations.

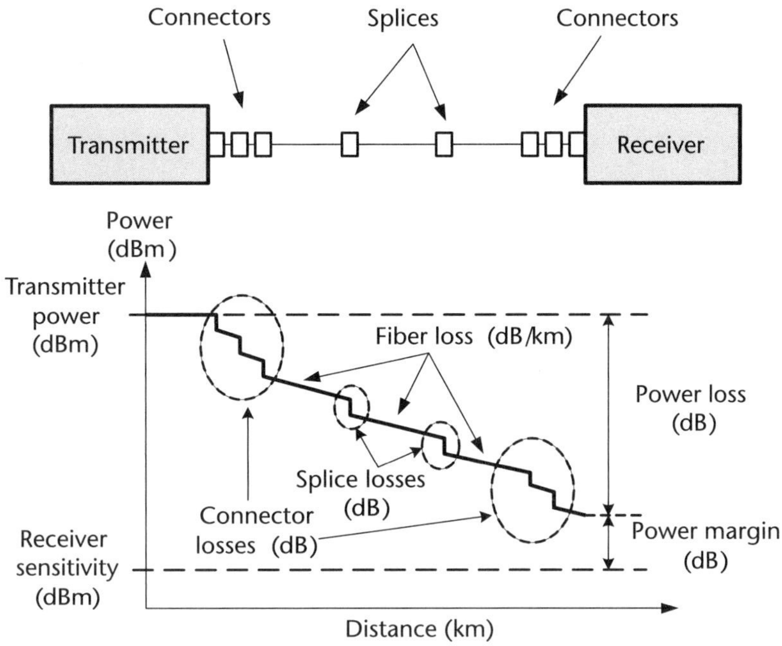

Figure 3.9 Optical power budget.

From available power we must subtract out all of the losses including cable attenuation, connector losses, and splice losses. Cable attenuation is often the most significant loss factor and should be determined by getting the exact number off of the cable being installed or using the manufacturer's worst-case number of the type of fiber you planned to install.

Although small, there is no perfect loss-less splice. Correctly done, each *fusion splice* should not introduce more than 0.1 dB of additional loss, and the fiber installers should be able to provide a guaranteed worst-case value. Today, splice losses (for single-mode fibers) of 0.05 dB or less are very common.

Connectors are another source of power loss. Depending on the end-to-end network design and number of patch panels, a typical long haul installation will have three to six connectors in the installation (e.g., equipment to patch cord and patch cord to pigtail). Table 3.3 contains some typical values, which can be used to approximate optical link budget calculations. If at all possible, real numbers from the network in question should be used.

Table 3.3 Fiber-Optic Losses

TIA standard for connector loss (many connectors have loss of less than 0.2 dB)	0.75 dB
Typical cable attenuation at 850 nm (MM)	3.0 dB
Typical cable attenuation at 1,310 nm (MM)	1.0 dB
Typical cable attenuation at 1,310 nm (SM)	0.4 dB
Typical cable attenuation at 1,550 nm (SM)	0.2 dB
Typical distance between splices	6.0 km
Typical system margin	3.0–6.0 dB
Typical splice attenuation (usually < 0.05 dB for SM and <0.1 for MM)	0.1 dB

Each of these losses, cable attenuation, connector loss, and splice loss is then subtracted from the available power. If the result is negative, there is no need to continue, as there is not enough power to drive the network.

An estimation of the number of anticipated repairs over the life of the fiber needs to be made. These repairs will add splice loss, so the number of anticipated splices is multiplied by the loss of each splice (same number used above), and subtracted from the remaining power.

Finally, we must account for temperature extremes, as well as any other unforeseen factors; for example, lasers and LEDs in the transmitter may age and lose power; connectors or splices may degrade; or connectors may get dirty if opened for rerouting or testing. Determining a *system margin* will be different for every organization depending on how much risk they want to assume in their network. Typically, a value around 3–6 dB is used, with longer fibers requiring higher system margin. The simple worksheet in Table 3.4 can help in these calculations.

It is always a good engineering practice to check the *maximum allowed receiver signal* (from a manufacturer's specifications) because input signal that is too high could produce intolerably high BER and even damage the receiver circuitry.

3.4.10 Testing Fiber-Optic Systems

Optical Time Domain Reflectometer (OTDR) is widely used in the telecommunications industry for testing bare and cabled fiber, including final fiber-optic link commissioning. OTDR can measure the attenuation coefficient of fiber and is useful to analyze discreet events in a link such as splice points or connector pairs. In addition, OTDR is very useful in locating damaged or distressed cable or broken fibers. More information on testing fiber-optic systems is provided in Chapter 7 [11].

Table 3.4 Optical Budget Calculator

Optical Budget Calculator	
Minimum transmit power	–10.0 dBm
Minimum receive sensitivity	–33.0 dBm
Available power	23.0 dB
(20 km of cable) × (0.25 dB/km)	5.0 dB
(6 connectors) × (0.75 dB/con)	4.5 dB
(4 splices) × (0.1 dB/splice)	0.4 dB
Total losses	9.9 dB
Link loss margin	13.1 dB
(5 repair splices) × (0.1 dB/splice)	0.5 dB
System margin	3.0 dB
Excess power	9.6 dB

3.5 Review Questions and Problems

1. Discuss the historic background of English and SI units of measurement and explain the importance of both when it comes to communications engineering.
   A. Which units are repeatable, and can be tested and proven in the lab environment? Research and compare gram/kilogram and pound as an example.
   B. Which system is better suited for science and engineering and why?
2. List some of the main differences between multimode and single-mode optical fibers. What is their area of application and why?
3. Explain the role of basic components of the 200-mile long fiber-optic system. How many optical regenerators do you need for this system?
4. What strategic locations would you select for regenerators along the long OPGW system? Think of such factors as access and maintenance, AC/DC power supply, and security.
5. Convert 0.25 dB/km into decibels per mile. (Answer: approximately 0.40 dB/mi)
6. The transmission capacity of a fiber depends on the fiber length. The longer a fiber is, the more detrimental certain effects such intermodal or chromatic dispersion are, and the lower is the achievable transmission rate. As mentioned earlier in this chapter, PMD is especially difficult to deal with. Calculate the maximum transmission length of a 40-Gbps fiber-optic system with fibers having a PMD coefficient of 0.1 $\frac{ps}{\sqrt{km}}$.
   Solution:

$$L = \frac{10^4}{B^2 \cdot PDM_{coeff}^2} = \frac{10^4}{40^2 \cdot 0.1^2} = 625 \text{ km}(338 \text{ mi})$$

   Discussion:

   A 40 Gbps fiber-optic system, with fibers having PMD coefficient of 0.1 $\frac{ps}{\sqrt{km}}$, will have a maximum transmission length of approximately 625 km (388 miles).
   A. For a fiber described above, calculate the maximum capacity of a fiber-optic system that is 1,000-miles-long. (Answer: ≈ 25 Gbps.)
   B. Discuss how the maximum allowed transmission length is changing with the increase in channel capacity.
7. Calculate excess power for a 30-mile-long fiber-optic system with a total of seven splices and six connectors, a minimum transmit power of -8 dBm, a minimum receive sensitivity of –34 dBm, and 6 dB of system margin. Fiber has losses of 0.23 dB/km, and five repair splices are planned for the future. (Answer: 3.2 dB)

8. List some of the main IEEE and ITU standards and/or recommendations that describe construction of fiber-optic cables for different applications.
9. A regenerator's enclosure and small fiber splice enclosures could become a target for practice shooting vandals. What are the ballistic requirements for the outdoor telecommunication enclosures, cabinets, and shelters in the country/state/area where you live?
10. Explain why SONET is still a preferred technology when it comes to power utility fiber-optic communications systems. Is SONET going to be replaced with some type of packet-based network (Ethernet and/or MPLS) in the near future, and what are the challenges in doing so?

References

[1] ITU-T G.652, *"Characteristics of the Single-mode Optical Fiber Cable,"* 2009.

[2] Proakis, G. J., *Willey Encyclopedia of Telecommunications*, Volume 3, Willey-Interscience, Hoboken, New Jersey: John Wiley and Sons, 2003.

[3] Allen, C., et al., "PMD Characterization on an Active Fiber Link: Final Report," The University of Kansas, sponsored by Sprint Corporation, 2003

[4] Lehpamer, H., *Transmission Systems Design Handbook for Wireless Networks,* Norwood, MA: Artech House, 2002.

[5] IEEE Std 1615-2007, "IEEE Recommended Practice for Network Communication in Electric Power Substations," 2007.

[6] IEEE Std. 1138-2009, IEEE Standard for Testing and Performance for Optical Ground Wire (OPGW) for Use on Electric Utility Power Lines.

[7] Commscope, *Fiber Optic Cable Products Broadband Applications & Construction Manual,* 2014

[8] IEEE Std. 1222-2011, IEEE Standard for Testing and Performance for All-Dielectric Self-Supporting (ADSS) Fiber Optic Cable for Use on Electric Utility Power Lines, 2011.

[9] Takaaki, O., "Recent Status and Trends in Optical Submarine Cable Systems," *NEC Technical Journal,* Vol. 5, No. 1, 2010.

[10] http://www.thefoa.org/ (accessed May 14, 2015).

[11] Anderson, R. D., et al., *Troubleshooting Optical Fiber Networks: Understanding and Using Optical Time-Domain Reflectometers, Second Edition,* Elsevier Academic Press, 2004.

CHAPTER 4

Introduction to Radio System Design

4.1 Radio Systems and Frequency Spectrum

This section will describe the basics of radio propagation, radio spectrum, and terminology used to describe electromagnetic waves.

4.1.1 Maxwell's Equations

James Clerk Maxwell (1831–1879) was a Scottish physicist and mathematician whose major discovery of the ether described the vast sea of space that made possible the transmission of light, heat, and radio waves. Maxwell's discovery of the ether (or the metaphor of it) led to many advances in electronic communications. Maxwell is regarded as the greatest theoretical physicist of the 19th century, and his extension of the electromagnetic theory of light led directly to Heinrich Hertz's discovery of radio waves and to the related advances in science and technology of today.

Maxwell's mathematical equations,[1] expressing the behavior of electric and magnetic fields and their interrelated nature, were valid, even though his theory of the ether was not. His calculations were scientific observations resulting in his conclusion that the speed of propagation of an electromagnetic field is approximately that of the speed of light (3×10^8m/sec) Maxwell's proposal that the phenomenon of light is therefore an electromagnetic phenomenon seemed to fit what he and other scientists could observe about the world around them. Maxwell concluded that visible light forms only a small part of the entire spectrum of possible *electromagnetic radiation* (EMR).

Maxwell's work of 1873, *The Complete Laws of Electrodynamics,* defines the relationship between the electric field quantities and the magnetic field quantities, and although a detailed explanation of these laws is beyond the scope of this book, they deserve to be mentioned here at least briefly. Maxwell's classic text was the

1. It was the physicist Ludwig Boltzmann who said that "There is nothing more practical than a good theory," and that is absolutely true in case of Maxwell's equations. Maxwell's equations beautifully and correctly describe both the energy storage field and the energy propagation field. Although highly theoretical and mathematical in their nature, they provided an unprecedented contribution to our understanding of the world around us as well as very practical side in the form of the development of radio communications

first to correctly assemble the complete laws of electrodynamics. Modern electromagnetism theory is based on the four fundamental equations known as Maxwell's equations.

Before Maxwell, the laws of electrodynamics, including *Gauss's Law, Ampere's Law of Magnetostatics,* and *Faraday's Law,* were laws of electrostatics, and they did not predict waves. These laws correctly described what is known as the near field (that is, the electrostatic field of an electric charge and the magnetostatic field of a current loop).

The earlier laws described the observable impact of electric charges and magnetic fields close to the source but failed to describe the distant impact of these forces. In the static case, when all electric charges are permanently fixed or if they all move at a steady state, the electric field and the magnetic field are not interconnected. This allows us to study electricity and magnetism as two distinct and separate phenomena.

Maxwell corrected and renamed Ampere's law of magnetostatics to become Ampere's law, so that consistency with the *law of conservation of charge* occurred. Maxwell added a term indicating that vortices of magnetic fields can be displacement current density (time-varying electric flux density) as well as conduction current density. The resulting corrected equations define the complete laws of electrodynamics and predict electromagnetic waves. Heinrich Rudolf Hertz confirmed experimentally that these waves exist.[2]

The latest developments of quantum physics have shown that Maxwell's equations are an approximation, albeit a very good one, and for most practical applications they will suffice.

4.1.2 Radio Propagation

Radio waves and microwaves are forms of electromagnetic energy that are collectively described with the term radio frequency. RF emissions and associated phenomena can be discussed in terms of energy, radiation, or fields. We can define radiation as the propagation of energy through space in the form of waves or particles [1].

Electromagnetic radiation can best be described as waves of electric and magnetic energy moving together (i.e., radiating) through space as illustrated in Figure 4.1. Electromagnetic waves are transverse (i.e., the magnetic field is perpendicular to the electric field and their cross-product points towards the direction of propagation).

The vector $\vec{P}$ in the direction of propagation is called the *propagation vector* or *Poynting vector.*[3]

$$\vec{P} = E \times H \tag{4.1}$$

2. Heinrich Hertz (1857–1894) was a German physicist who applied Maxwell's theory to the production and reception of radio waves. In recognition of his work, the unit of frequency of a radio wave—one cycle per second—is named the hertz.

3. It is named after the English physicist John Henry Poynting, who introduced it in 1884.

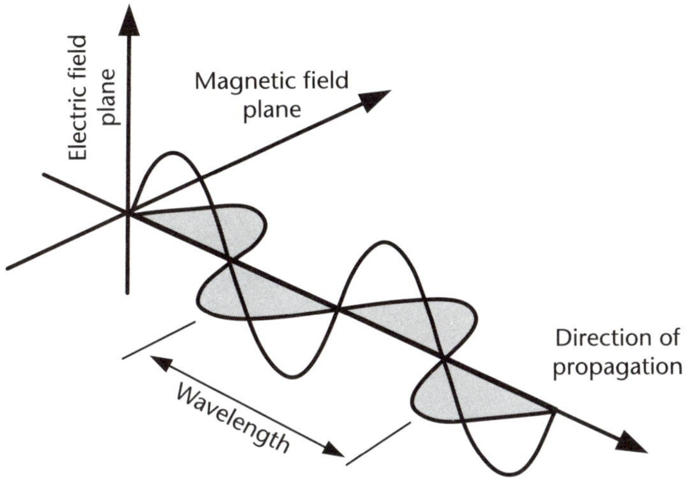

Figure 4.1 Electromagnetic wave.

The existence of propagating electromagnetic waves can be predicted as a direct consequence of Maxwell's equations. These equations specify the relationships between the variations of the vector electric field, E, and the vector magnetic field, H, in time and space within a medium. The E-field strength is measured in volts per meter and is generated by either a time-varying magnetic field or a free charge. The H-field is measured in amperes per meter and is generated by either a time-varying electric field or a current.

Maxwell's first two equations contain constants of proportionality that dictate the strengths of the fields. These are the permeability of the medium, μ, in henrys per meter and the permittivity of the medium, ε, in farads per meter. They are normally expressed relative to the values in free space:

$$\begin{aligned}\mu &= \mu_0\mu_r \\ \varepsilon &= \varepsilon_0\varepsilon_r\end{aligned} \tag{4.2}$$

where $\mu_0 = 4\pi \times 10^{-7}$ H/m and $\varepsilon_0 = 8.854 \times 10^{-12}$ are the values in free space, and μ_r and ε_r are the relative values (i.e., $\varepsilon_r = \mu_r = 1$ in free space and greater elsewhere). Free space strictly indicates a vacuum, but the same values can be used as good approximations for dry air at typical temperatures and pressures.

Maxwell calculated that electromagnetic waves would propagate at a speed given by the equation:

$$c = \frac{1}{\sqrt{\mu_0\varepsilon_0}} \tag{4.3}$$

which is actually the speed of light.[4]

4. Actually, this relationship had been discovered by Wilhelm Eduard Weber and Rudolf Kohlrausch prior to the development of Maxwell's electrodynamics; however Maxwell was the first to produce a field theory

The term *electromagnetic field* is used to indicate the presence of electromagnetic energy at a given location. The RF field can be described in terms of the electric and/or magnetic field strength at that location. Like any wave-related phenomenon, electromagnetic energy can be characterized by a frequency and a wavelength.

4.1.3 Frequency and Wavelength

Frequency (f) is defined as a number of cycles, or periods, per unit of time, and measured in hertz. The wavelength (λ) of a sinusoidal wave is the spatial period of the wave—the distance over which the wave's shape repeats.

In other words, the wavelength λ is the distance by which the phase of the sinusoidal wave changes by 2π radians (by definition, $k\lambda = 2\pi$, $\omega = kc$, and $f = \omega/2\pi$). Electromagnetic waves travel through free space at the speed of light, and the wavelength and frequency of an electromagnetic wave are inversely related by a simple mathematical formula connecting wavelength, speed of light (c), and frequency.

$$\lambda = \frac{2\pi}{k} = \frac{2\pi c}{\omega}$$
$$\lambda = \frac{c}{f} \tag{4.4}$$

We can see that the high-frequency electromagnetic waves have short wavelengths and that low-frequency waves have long wavelengths. Variation, or *modulation,* of the properties of the wave (amplitude, frequency, or phase) then allows information to be carried in the wave between its source (transmitter) and destination (receiver), which is the goal of radio communications.

4.1.4 Polarization

A very important parameter of the electromagnetic wave is the polarization, which can be described by the locus of the tip of the electric field vector, E, in time. In a trigonometric form, a wave propagating in the z direction can be expressed as follows:

$$\vec{E} = \hat{x}A\cos(\omega t - Bz) + \hat{y}B\sin(\omega t - Bz) \tag{4.5}$$

Here, A and B are the amplitudes of the field components in the x and y directions, respectively. If $A = 0$ or $B = 0$, we get a linearly (either vertically or horizontally) polarized wave. If $A \neq 0$ and $B \neq 0$, we have an elliptically polarized wave. If $A = B$, the result is a *circularly polarized wave,* widely used in satellite communications.

consistent with waves traveling at the speed of light. Maxwell concluded that light is an electromagnetic wave having wavelengths that it can be detected by the eye. Experimental verification came within a few years, but not before Maxwell's death.

One of the reasons for the use of circularly polarized waves in satellite communications is that the ionosphere causes *Faraday rotation*[5] to electromagnetic waves and therefore a linearly polarized electromagnetic wave may be rotated by an unknown and unpredictable amount. This phenomenon does not represent a problem for circularly polarized waves.

Some antennas, such as the helical antenna, produce circular polarization; however, circular polarization can be generated using two linearly polarized antennas by feeding them with an equal magnitude signal but with a 90 phase shift between them.

Generally speaking, any two antennas that form a link with each other (transmitter and receiver) must be set for the same *polarization.*

4.1.5 Electromagnetic Spectrum

The electromagnetic spectrum is separated into many categories and subcategories, based on the frequency and wavelength, source, and applications of the electromagnetic waves. The electromagnetic spectrum[6] includes all the various forms of electromagnetic energy (see Figure 4.2).

Different frequencies have different propagation characteristics. All frequencies are attenuated and reflected by materials to a greater or lesser degree, with the higher frequencies being more attenuated than the lower frequencies. This allows lower frequencies to have significant signal-penetration capabilities through all materials including metal.

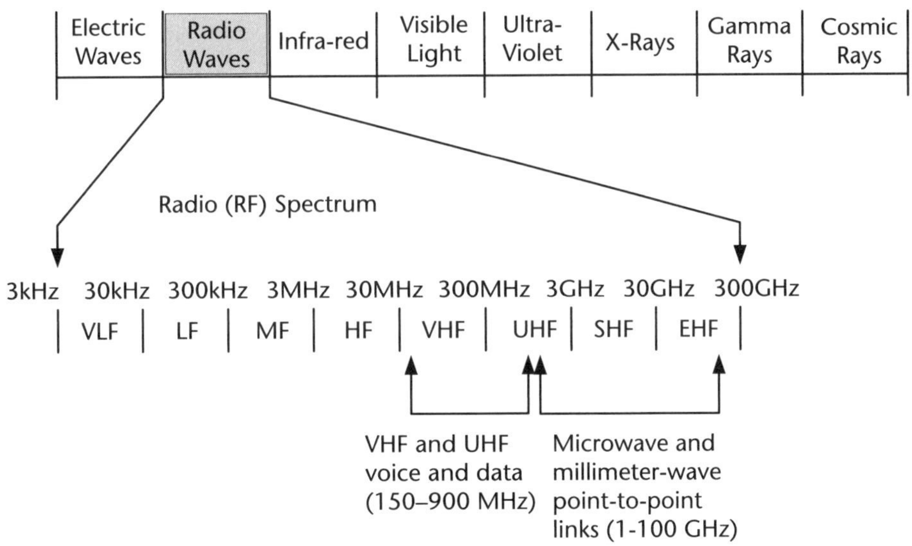

Figure 4.2 Electromagnetic spectrum.

5. Discovered in physics by Michael Faraday in 1845, the Faraday Effect (or Faraday rotation) was the first experimental evidence that light and electromagnetism are related. Satellite systems like GPS have utilized circular (right-hand) polarization, and not linear polarization for transmission to ground receivers. The Faraday rotation is higher at lower frequencies but declines with the frequency.

6. The word spectrum literally means a range of values or a set of related quantities.

Radio waves are divided into many types, depending on their applications, ranging up to microwaves and millimeter waves at their highest frequencies. The radio part of the electromagnetic spectrum is generally defined as that part of the spectrum where electromagnetic waves have frequencies in the range of about 3 kHz–300 GHz.

Infrared radiation lies below visible light in frequency and is produced by thermal motion and the vibration and rotation of atoms and molecules. Infrared's lower frequencies overlap with the highest-frequency microwaves.

Visible light is largely produced by electronic transitions in atoms and molecules, and is defined as being detectable by the human eye. Its colors vary with frequency, from red at the lowest to violet at the highest.

Ultraviolet radiation starts with frequencies just above violet in the visible range and is produced primarily by electronic transitions in atoms and molecules.

X-rays[7] are created in high-voltage discharges and by electron bombardment of metal targets. Their lowest frequencies overlap the ultraviolet range but extend toward much higher values, overlapping at the high end with gamma rays.

Gamma rays are nuclear in origin and are defined to include the highest-frequency electromagnetic radiation of any type.

4.2 Antenna Parameters

An antenna is a device that transmits and/or receives electromagnetic waves. Antennas demonstrate a property known as reciprocity, which means that an antenna will maintain the same characteristics, including the radiation pattern whether it is transmitting or receiving.

Most antennas are resonant devices operating efficiently over a relatively narrow frequency band.

4.2.1 Radiation Pattern

Radiation patterns describe the distribution in space of electromagnetic energy generated (or received) by a given antenna. For mobile, portable, and some base station applications, the type of antenna needed has an omnidirectional radiation pattern. The omnidirectional antenna radiates and receives equally well in all horizontal directions, while directional antennas focus energy in a particular direction. (See Figure 4.3.) For microwave point-to-point links, required antennas are always highly directive.

The antenna radiation lobe containing the maximum power (exhibiting the greatest field strength) is called the *main lobe* (or the *main beam*). The horizontal radiation pattern is plotted as a function of azimuth around the antenna and is usually specified. The width of the main lobe is usually specified as the angle encompassed between the points where the power has fallen 3 dB below the maxi-

7. X-radiation is referred to with terms meaning Röntgen radiation, after Wilhelm Röntgen, who is usually credited as its discoverer in 1895, and who had named it X-radiation to signify an unknown type of radiation. In 1901 Röntgen received a first Nobel Prize for his discovery.

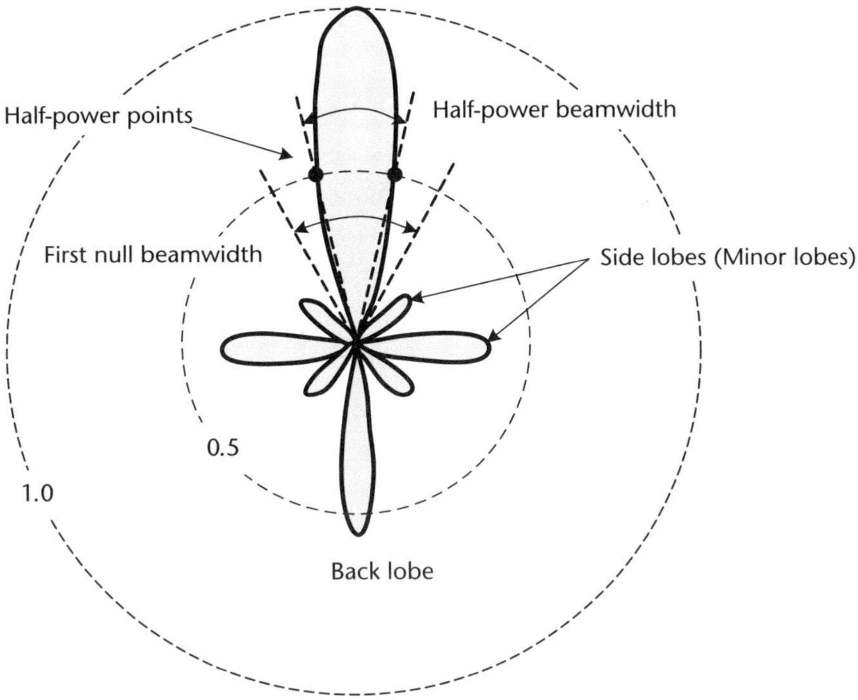

Figure 4.3 Antenna radiation pattern.

mum value. The vertical radiation pattern, plotted as a function of elevation from a specified azimuth, is also of interest, and may be similarly specified.

Sidelobes are the radiation lobes in any direction other than that of the main lobe.

4.2.2 Beamwidth and Gain

The half-power beamwidth is the angular separation between the half-power points on the main beam antenna radiation pattern, where the gain is one-half the maximum value (–3 dB).[8] An antenna's gain and radiation pattern are fundamentally related; higher-gain antennas always have narrower beamwidth (patterns), and low-gain antennas always have wider beamwidth.

The *directive gain* is a measure of the antenna ability to direct its power toward a given direction. Engineers commonly refer to *half-wave dipole* antenna gains. Compared to the gain of an "ideal" *isotropic antenna* of 1 (0 dB), the gain of a half-wave dipole antenna is 1.64 (2.15 dBi). The relationship between the two units, decibels relative to an isotropic radiator, and decibels relative to a dipole radiator, is therefore expressed as follows:

$$G\ (\text{dBi}) = G\ (\text{dBd}) + 2.15 \tag{4.6}$$

8. Beamwidth is usually measured at the half-power or –3-dB point unless otherwise specified. A 3-dB beamwidth is approximately equal to the angle from the peak of the power to the first null.

An isotropic radiator is a hypothetical, lossless antenna having equal radiation in all directions. The three-dimensional radiation pattern of an isotropic radiator is a sphere. Considering both stations of a radio link, the difference between free-space loss using isotropic and half-wave dipole antennas is about 4.3 dB.

When antenna manufacturers simply specify the gain of an antenna, they are usually referring to the maximum value of gain, G. Antenna gain and other specifications are defined and measured in the far field of the antenna.

In an antenna radiation pattern, a *null* is a zone in which the effective radiated power is at a minimum. A null often has a narrow directivity angle compared to that of the main beam. Thus, the null is useful for several purposes, such as suppression of interfering signals in a given direction.

4.2.3 Front-to-Back Ratio

Minimizing RF interference is one of the toughest challenges for radio communications network designers. One of the main causes of interference between radio links is the existence of sidelobes [a radiation in any direction other than the direction(s) of intended radiation] and back lobes (the radiation lobe opposite to the main lobe), spurious reflections that radiate (and receive) in directions other than the main beam. Most of the sidelobes are a result of imperfect antenna design and construction.

Front-to-back ratio is expressed in decibels, and as the name implies, it is the ratio of the maximum signal in the forward direction to the signal in the opposite direction.

4.2.4 Bandwidth

The bandwidth of an antenna expresses its ability to operate over a wide frequency range. It is often defined as the range over which the power gain is maintained to within 3 dB of its maximum value, or the range over which the VSWR is no greater than 2:1, whichever is smaller. The radiation pattern of an antenna may change dramatically outside its specified operating bandwidth.

The bandwidth can be described in terms of percentage of the center frequency of the band.

$$BW = 100 \times \frac{f_H - f_L}{f_C} \ [\%] \tag{4.7}$$

where f_H is the highest frequency in the band, f_L is the lowest frequency in the band, and f_C is the center frequency in the band. In this way, bandwidth is constant relative to frequency. If bandwidth was expressed in absolute units of frequency, it would have different values, depending upon the center frequency. Different types of antennas have different bandwidth limitations.

4.2.5 Antenna Selection

The selection of the correct antenna should be based on the application as well as the results of radio path analysis and radio propagation calculations. The antenna type and size must be determined before a frequency coordination study can be performed and before applying for the license to operate the radio system [2].

Generally speaking, the main parameters of interest when choosing any type of antenna are operating frequency band, radiation pattern, gain, and half-power beamwidth (degrees).

Selection of a microwave parabolic dish antenna is more complex; some additional parameters of interest when choosing a parabolic dish antenna are polarization (single or dual-polarized), wind load, front-to-back ratio, and cross-polarization discrimination and a large number of additional options that may be ordered from manufacturers.

There are literally hundreds of different special-application antennas in use today. Antennas can be broadly classified as *omnidirectional* and *directional antennas*, which depends on the directionality. Probably two most commonly used types of directional antennas of interest here are microwave reflector (parabolic dish) and Yagi-Uda antennas.

4.2.5.1 Omnidirectional Antennas

In radio communication, an omnidirectional antenna (or a so-called omni antenna) is a class of antenna that radiates uniformly in all directions. Common types of low-gain omnidirectional antennas include the whip antenna, "rubber duck" antenna, and ground plane antenna. Omnidirectional radiation patterns are produced by the simplest practical antennas, monopole and dipole antennas, consisting of one or two straight rod conductors on a common axis.

Unlike isotropic antennas, dipole antennas are real antennas. The dipole radiation pattern is 360 degrees in the horizontal plane and approximately 75 degrees in the vertical plane (this assumes the dipole antenna is standing vertically) and resembles a donut in shape. Because the beam is slightly concentrated, dipole antennas have a gain over isotropic antennas of 2.15 dB in the horizontal plane. Dipole antennas are said to have a gain of 2.15 dBi, which is in comparison to an isotropic antenna.

The higher the gain of the antennas, the smaller the vertical beamwidth is. Increasing the gain of omnidirectional antennas results in less energy sent vertically and more energy sent horizontally, extending the range.

Omnidirectional antennas mount vertically and transmit and receive equally in all directions within the horizontal plane. It is important to understand the lack of a signal very close to the antenna. Most dipole omni-antennas have a minimum distance for optimum signal reception.

4.2.5.2 Microwave Reflector Antennas

The microwave reflector antenna[9] is a highly directional, parabolic, dish-shaped radiator that is connected to the microwave transmitter (through the RF branching network—RF circulators, filters, couplers, and switches) via transmission lines (coaxial cable or waveguide). Applications include microwave link applications—satellite communication systems, navigation systems, terrestrial communications systems, deep space communication systems, and radio astronomy.

Antennas are available in many sizes to satisfy the requirements of a particular application—the larger the antenna diameter, the higher the antenna gain and smaller the beamwidth. For a parabolic reflector microwave antenna (above 1 GHz) consisting of a dish-shaped surface illuminated by a feed horn (feeder) mounted at the focus of the reflector, antenna gain, G, is approximately given as

$$G = 20\log(D) + 20\log(f) + 7.5 \quad [\text{dBi}] \tag{4.8}$$

where

dBi = decibels over an isotropic radiator;

D = antenna dish diameter in feet;

f = frequency (GHz).

Although not shown in this simplified formula, antenna gain is related to directivity through the *aperture efficiency* (η) parameter, expressed as a percentage or in decibels [3]. This formula for beamwidth assumes the efficiency of a parabolic antenna to be about 55 percent (or 0.55), although some manufacturers may be able to improve on this value, so the gain given by a manufacturer for a specific antenna should be used when available. In most other cases, the preceding formula will suffice.

The *beamwidth* of a parabolic antenna can be approximated by the following formula:

$$BW \approx \frac{70}{f \cdot D_{feet}} \quad [\text{deg}] \tag{4.9}$$

where

f = frequency (GHz);

D = diameter of parabola in feet.

9. Initial design and experimentation with microwave antennas began more than 100 years ago, and a number of microwave systems using parabolic antennas grew significantly during the 1930s. During World War II, designs such as pencil-beam and shaped-beam antennas were developed for radar systems used by the Allies. While many advances were made at that time, it was in the 1950s that terrestrial microwave communication systems were deployed, and parabolic reflector designs were utilized on the commercial systems.

The small horn antenna is commonly used as a feed to a parabolic dish antenna, a gain standard for antenna gain measurements, and as compact medium-gain antennas for various systems. The gain of the horn antenna can be calculated to within 0.1 dB accuracy from its known dimensions and is therefore used as a gain standard in antenna measurements.

To make the dish lighter and to withstand strong wind, a dish made of metal mesh instead of solid metal can be used.[10] These grid-type dish antennas are usually employed for the low-frequency microwave bands at 1.4 and 2 GHz.

4.2.5.3 Yagi-Uda Antennas

A Yagi-Uda antenna (commonly known simply as a Yagi antenna)[11] is a directional antenna used at frequencies from 10 MHz to 3 GHz and consists of a single driven dipole with a reflector dipole behind the driven element and one or more parasitic director elements in front; a common design uses one reflector and from one to seven director elements (see Figure 4.4).

As the number of elements is increased, directivity and gain increase. Bandwidth, however, decreases as the number of elements is increased; therefore, arrays of more than four director elements are typically considered narrowband.The radiating element is a half-wavelength at the center of the band covered. The single reflector element is slightly longer, and the director elements are slightly shorter, all spaced approximately a quarter-wavelength from each other.

Table 4.1 shows how the number of elements determines the gain and beamwidth of a Yagi-Uda antenna. To achieve the same gain at lower frequencies, large antenna size becomes a challenge, so for practical reasons this antenna is typically used for frequencies above 100 MHz, where the antenna dimensions are manageable.

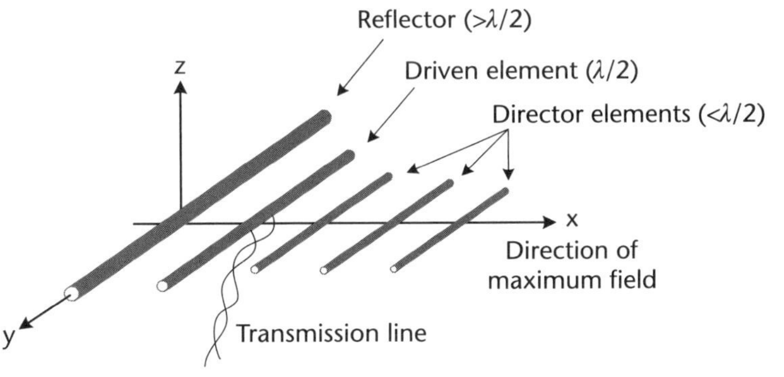

Figure 4.4 The Yagi-Uda antenna.

10. There are a number of other types of reflectors but they are not used in microwave point-to-point communications. For example, at low microwave frequencies or UHFs, a parabolic dish becomes big, so only a portion of the dish is used instead. This is called a truncated parabolic dish (also known as an orange peel paraboloid reflector), commonly used on ships.

11. This antenna was invented in 1926 by Shintaro Uda of Tohoku Imperial University, Japan, with a lesser role played by his colleague Hidetsugu Yagi. However, the "Yagi" name has become more familiar (Yagi translated the original work from Japanese to English) with the name of Uda often omitted.

Table 4.1 Typical Characteristics of Yagi-Uda Antennas

Number of Director Elements	*Gain [decibels]*	*Beamwidth [degrees]*
2	3–4	65
3	6–8	55
4	7–10	50
5	9–11	45
9	12–14	37
15	14–16	30

There are no simple formulas for designing Yagi-Uda antennas[12] due to the complex relationships between physical parameters such as element length, spacing, and diameter and performance characteristics such as gain and input impedance [4]. Consequently, these antennas are often empirical designs using an element of trial and error, often starting with an existing design modified according to designers' experience and hunch.

4.3 Microwave Point-to-Point Systems

Terrestrial microwave point-to-point link is LoS radio communication technology that uses high-frequency beams of radio waves to provide high-speed radio connections that can send and receive voice, video, and data information.

Microwave links are widely used for point-to-point communications because their small wavelength allows small antennas to direct them in narrow beams, which can be pointed directly at the receiving antenna. This allows nearby microwave equipment to use the same (or very close) frequencies without interfering with each other, as lower frequency radio waves do.

4.3.1 Historic Developments

After experiments in the 1930s, the first commercial long-haul, 4-GHz multihop microwave point-to-point system finally entered operation in 1947, providing telephone and television circuits between New York and Boston. The eight-links route is: New York—Jackie Jones—Birch Hill—Spindle Hill—Glastonbury—Bald Hill—Asnebunskit—Bear Hill—Boston.

The two terminal points of this microwave radio relay system are the headquarters building of the American Telephone and Telegraph (AT&T) Company's long lines department at 32 Avenue of the Americas in New York and in Boston, at the Bowdoin Square building of the New England Telephone and Telegraph

12. A very similar configuration to the Yagi–Uda antenna is the log-periodic antenna. It produces a similar end-fire radiation pattern and directivity (typically between 7 and 15 dBi) and it is widely used in the VHF and UHF bands. The main difference is that log-periodic antenna has a much wider bandwidth than Yagi-Uda antenna.

Company. Between terminal points, the microwave system has eight links via seven intermediate repeaters spaced about 30 miles apart [5].

Network traffic for the next 35 years was carried primarily by a mixture of microwave relay and coaxial cable. The details of this first commercial microwave system were described in *Popular Science* magazine in 1947 [6]. Figure 4.5 shows what the microwave repeater site looked like in 1947.

A good overview of important discoveries in the development and applications of microwave technology from the time of Hertz until 1980 is presented in [7].

4.3.2 Microwave Radio

Today, microwave radios are deployed in the existing mobile radio infrastructures and in public transport organizations, railroads, power utilities, and other public utilities that are major users of microwave systems. Governments in many countries use fixed services to provide a number of defense and nondefense functions in locations and circumstances where commercial communications services do not satisfy requirements. They use microwave point-to-point services for control and monitoring of many wide-area systems (e.g., air traffic control) for connecting government mobile radio sites, for tactical communications, and for communications within test and training ranges [8].

Most of the commercially used terrestrial microwave point-to-point (also called *radio-relay*) systems use frequencies of approximately 2–90 GHz. Radio links may be established between any two points within the LoS and, depending on the frequency, geographical region, and rain statistics, the typical link distance can be up to 25 mi (40 km), although much longer links are not uncommon either. For the longer microwave link hops, additional measures have to be taken to ensure the required reliability of the system (e.g., space and/or frequency diversity).

A typical digital microwave radio consists of three basic components: a *digital modem* for interfacing with digital terminal equipment, a *RF unit* for converting a carrier signal from the modem to a radio signal, and an *antenna* used to transmit and receive the signal. The combination of these three components is called a *radio terminal.* Two terminals are required to establish a microwave communications link, commonly referred to as a *microwave hop* or *microwave link.*

Figure 4.5 AT&T microwave repeater site in 1947.

From a construction point of view, microwave radios come in three different configurations (Figure 4.6)—all-indoor, split configuration, and all-outdoor.[13] From a design perspective, split configuration does not use high-loss waveguide (especially important at higher frequencies) and can provide longer microwave links. From a maintenance perspective, although it depends on circumstances, the all-indoor configuration is generally considered to be the most reliable and easiest to maintain.

Microwave point-to-point communication can be achieved by a single connection—for example, a microwave link between two stations located at specified fixed points or multiple cascaded (daisy-chained) links made by a number of intermediate repeaters with or without partial payload *drop-insert.*[14]

We can feed the data and voice traffic into the radio using an electrical or optical interface. In the radio, the digital signals are coded into analog signals and converted to microwaves (with a typical wavelength of a few centimeters). The microwaves are sent using a highly directive parabolic antenna. At the other end, the signals are received and restored to the digital format.

Microwaves, which are only centimeters (or inches) in length, are small relative to the surroundings and hence do not have the bending property. Therefore, to establish a radio link, it is important to have radio LoS between the two radio position sites. LoS exists when there is a direct path between two separate points and no obstructions (e.g., buildings, trees, hills, or mountains) between them.

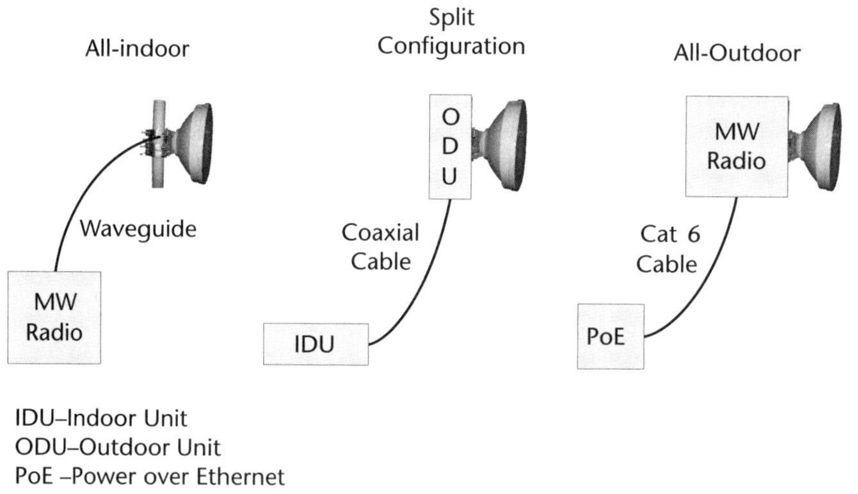

Figure 4.6 Three types of microwave radio configurations.

13. All-outdoor radio configurations today utilize Power over Ethernet (PoE) where power passes from the source to the radio over standard Ethernet CAT-5/6 cables (IEEE 802.3at or IEEE 802.3af). There are four twisted pairs in each CAT-5/6 cable, so Ethernet signals travel along two twisted pairs, one pair for each direction, and power uses one spare pair for the positive DC supply and the other spare pair for the negative return. Distance is limited to about 100m (330 ft). Another method involves actually using a pair of wires that's already being used to carry data.

14. Drop-insert is a functionality provided in analog and digital repeaters, where only radio-system specific control and service channels, and possibly part of the payload, is separated and made available for local traffic and system management and maintenance.

Unobstructed paths are always preferred, because the outage times and percent reliability computed for obstructed paths may not be reliable.

One or more radio paths connected in tandem form a *microwave system*. The radio stations between two terminal stations are called *repeater stations* (active or passive).

Repeaters can be *nonregenerative* when the signal is only filtered and amplified, with or without down and up conversions (e.g., in some older analog FDM systems) or *regenerative* when, in digital applications, the signal is demodulated and remodulated before transmission to the next radio hop.

Passive repeaters implemented without any active radio components (e.g., two directional antennas connected back to back or a large size reflector) are also utilized. Aside from a clear Fresnel zone, these systems have a number of other special requirements in order to make passive repeaters work properly.

4.3.3 Radio Propagation Theory

4.3.3.1 Introduction

While signals in lower-frequency bands can propagate for many miles and penetrate more easily through buildings, microwave signals have more limited range, while millimeter-wave signals can travel only a few miles or less and do not penetrate solid materials very well. Millimeter-wave propagation[15] has its own peculiarities but the millimeter-wave spectrum is of increasing interest because of the wide bandwidths available for carrying communications at this frequency range, valuable in supporting applications such as high-speed data transmission and video distribution.

Figure 4.7 illustrates the radio frequency propagation mechanisms for microwave- and millimeter-wave frequencies including diffraction, refraction, reflection, and scattering.

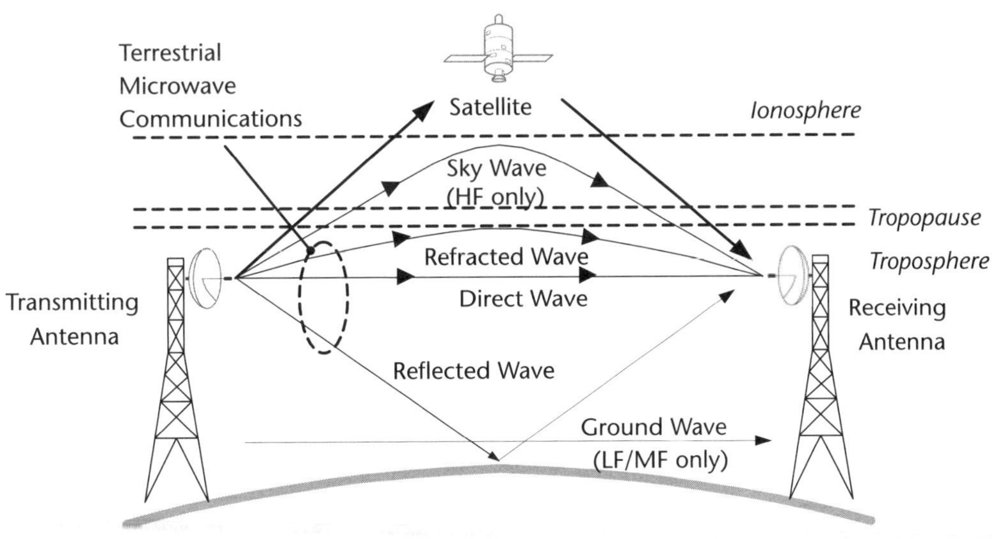

Figure 4.7 Radio wave propagation.

15. According to the ITU, electromagnetic waves between 30 and 300 GHz are called millimeter waves (MMW) instead of microwaves (MW) because the wavelengths for these frequencies are about 1–10 mm.

4.3.3.2 Standard Radio Propagation Mechanisms

Standard propagation mechanisms are those mechanisms and processes that occur in the presence of a standard atmosphere. These propagation mechanisms are free-space propagation, reflection, diffraction, scattering, and tropospheric scatter.

The simplest case of electromagnetic wave propagation is the transmission of a wave between a transmitter and a receiver in a *homogeneous atmosphere* (sometimes also called *free space*). It is defined as a region whose properties are isotropic, homogeneous, and loss-free (i.e., away from the influences of the Earth's atmosphere).

Reflection occurs when an electromagnetic wave strikes a nearly smooth, large surface, such as a water surface, and a portion of the energy is reflected from the surface and continues propagating along a path that defines an angle with the surface equal to that of the incident ray. Obstruction dimensions are very large compared to the signal wavelength [9]. The strength of the reflected wave is determined by the reflection coefficient, a value that depends on the frequency and polarization of radiation, the angle of incidence, and the roughness of the reflecting surface. Reflection rays from different surfaces may interfere constructively or destructively at a receiver causing *multipath propagation.*

Diffraction occurs when an impenetrable body obstructs the radio path between the transmitter and receiver. Energy tends to follow along the curved surface of an object. Diffraction is one of the important physical phenomena that have been studied using different approaches. Based on Huygens' principle, secondary waves are formed behind the obstructing body, and the study is usually mainly based on physical optics.

The ability of the electromagnetic wave to propagate beyond the horizon by diffraction is highly dependent on frequency; the lower the frequency, the more the wave is diffracted. Over-the-horizon communication systems, for example, are based on this principle.

Scattering occurs when the radio channel contains objects whose dimensions are approximately the same as or smaller than the propagating wavelength. Scattering, which follows the same physical principles as diffraction, causes energy to be radiated in many different directions.[16]

4.3.3.3 Anomalous Radio Propagation Mechanisms

Standard atmosphere represents the desired mode of radio propagation; other modes of propagation can seriously impact the terrestrial microwave links and their performance (Figure 4.8).

Experimental work in the late 1960s and early 1970s clearly identified a persistent meteorological phenomenon capable of confining, or ducting, microwave signals within the surface layer and propagating these signals to ranges well beyond the radio horizon.

16. Troposcatter communications are using small imperfections within the atmosphere's refractive structure to propagate well beyond the horizon. It was used by commercial services and the military 1950–1980, before the expansion of satellite services. It is still used today by military although mostly just as a backup communications system.

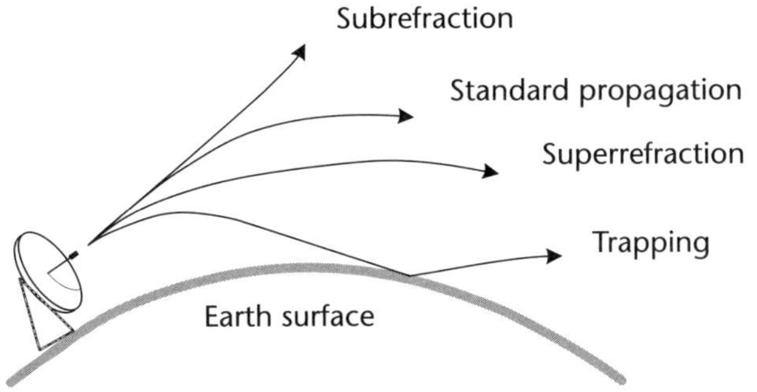

Figure 4.8 Types of radio propagation.

Anomalous meteorological conditions may occur that considerably change the standard propagation described above. A deviation from the normal atmospheric refractivity leads to conditions of anomalous propagation such as subrefraction, superrefraction, and trapping (ducting).

The altitude where ducting conditions occur over land is almost always within the lowest 200m of the atmosphere and caused by temperature inversions near the surface. Ducting events over land at polar latitudes and over deserts are almost entirely caused by the dry refractivity variability. Mid-latitude and tropical events appear at higher altitudes as well, up to 1.9 km and 2.3 km, respectively, while all events at higher altitudes are over the sea. [10, 11].

The meteorological phenomenon called an *evaporation duct* is related to the rapid decrease in water vapor (moisture) in the first few meters above the sea surface.[17]

The effects of anomalous propagation are hard to predict and almost impossible to completely avoid.

4.3.4 Effective Earth's Radius and k-Factor

Refraction is a physical phenomenon observed in any medium that has a varying refractive index, and it produces the effect of bending a light ray or microwave beam. Refraction in the atmosphere is described by its index of refraction, which is dependent on the humidity, temperature, and pressure of the atmosphere, all of which are a function of height [12].

In free space (ideal medium, vacuum for example), an electromagnetic wave will travel in a straight line, because the index of refraction is the same everywhere. Within the Earth's atmosphere, however, the velocity of the wave is less than in free space, and the index of refraction normally decreases with increasing altitude. Therefore, the propagating wave will be bent upward or downward from a straight line.

To facilitate path profiles at the time computers were not available, radio transmission engineers introduced the Earth-radius factor k to compensate for the

17. During World War II, A. S. Monin described the vertical distribution of meteorological quantities within the tropospheric surface layer (extending tens of meters vertically from the interface between the earth's surface and the troposphere). In the late 1960s, Monin-Obukhov Similarity (MOS) theory was experimentally confirmed over land and adapted to model microwave propagation over the sea.

refraction in the atmosphere. The effective Earth radius factor, k, is defined as the factor that is multiplied by the actual Earth radius, a, to give the effective Earth radius a_e. The mean Earth radius is 6,371 km.

Applying appropriate k-values to the true-Earth radius, an equivalent-Earth radius is geometrically obtained and, consequently, straight rays can be drawn.[18] Variations in atmospheric refractive conditions cause changes in the *effective Earth's radius* or *k-factor* from its median value of approximately 4/3 for a standard atmosphere. When the atmosphere is sufficiently subrefractive, the ray paths will be bent in such a way that the Earth appears to obstruct the direct path between transmitter and receiver, giving rise to the type of fading called *diffraction fading*. This fading is an important factor in determining the correct antenna heights.

The refraction varies considerably from day to day and from one place to another, and it is particularly variable over water. More information about refractivity can be found in [13].

4.4 Radio Path Engineering

4.4.1 Near and Far Fields

Maxwell showed that any conductor (e.g., antenna) supplied with an alternating current produces a varying magnetic field (H-field), which, in turn, produces electric field lines (E-field) in space. This is termed the near field. In the near field both the E and H fields are relatively static with no propagation. They only vary in strength as the current varies, with the magnetic flux of the H-field coming out from the antenna, and going back in, and the E-field emanating outward.

Maxwell also proved that beyond this quasistatic near field, both the E-fields and H-fields at a certain distance, detached themselves from the conductor and propagated into free space as a combined wave, moving at the speed of light with a constant ratio of $E/H = 120\pi$ or 377 Ω. The point in which this happens is called the *far field*.

These names imply that two regions, with a boundary between them, exist around an antenna, while, in fact, as many as three regions and two boundaries exist. These boundaries are not fixed in space, and they move closer to or farther from an antenna, depending on both the radiation frequency and the amount of phase error an application can tolerate.

When the distance is large, the spherical surface of uniform power density appears flat to a receiving antenna, which is very small compared to the surface of the sphere, so the *far field*[19] wave front is considered planar and the rays approximately parallel. In the far-field zone, electric field strength, magnetic field strength, and power density are interrelated by simple mathematical expressions, where any one of these parameters defines the remaining two. The beginning of the far field of an antenna with the largest dimension (D) in feet and wavelength (λ) in feet:

18. For distances of up to a few miles, the Earth can be treated as flat surface and k-factor really does not come to play.

19. Engineers perform link engineering, including Fresnel clearances and path profiles, based on the assumption that microwave antennas are in the far-field region.

$$R_{FF} = \frac{2D^2}{\lambda} \quad [\text{ft}] \tag{4.10}$$

The *far field,* or Rayleigh distance, is the region where the radiation pattern is independent of distance. In this area, fields vary with $1/r$. The far field is the most important because it determines the antenna's radiation pattern; it is also known as the *Fraunhofer region.*[20] It is important to notice that far-field expressions are valid if $D >> \lambda$.

4.4.2 Free-Space Loss

Electromagnetic waves are attenuated while propagating between two points. The free-space path loss model is used to predict received signal strength when the transmitter and receiver have a clear, unobstructed LoS path between them. The Friis free-space path loss model expressed here is valid only for distances that are in the far field of the transmitter antenna, and in the logarithmic (decibel) form, the equation is

$$L_{FSL} = 96.60 + 20\log\left(f_{GHz}\right) + 20\log\left(d_{mi}\right) \, [\text{dB}] \tag{4.11}$$

where

f = frequency (GHz);

d = LoS range between antennas (mi).

Free-space loss is always present, is directly proportional to the square of distance and frequency, and gives the total attenuation caused by wave propagation effects.

4.4.3 Fresnel Zones, Clearance Rules, and Path Profiles

Clear LoS is a preferred propagation mechanism in every radio network. Terrestrial microwave point-to-point systems operate under the assumption of a clear LoS and require an unobstructed first Fresnel[21] zone (see Figure 4.9). The concept of the Fresnel's ellipsoid is an integral part of the terrestrial microwave point-to-point link design and serves to check for obstructions that penetrate the ellipsoid and calculate a possible diffraction.

Natural obstacles, like terrain and trees, and man-made obstacles like buildings, wind farms [14], and electric transmission towers have to be avoided during the microwave link path design.

20. Joseph Fraunhofer (1787–1826) was a German optician, known for the discovery of the dark absorption lines known as Fraunhofer lines in the Sun's spectrum. He was a maker of excellent optical glass and achromatic telescope objectives, and invented extremely precise methods for measuring dispersion.

21. The early nineteenth century French physicist, Augustin Fresnel, made an important observation about the behavior of light. Fresnel noted that a ray of light passing near a solid object is subject to diffraction, or bending. This diffraction causes the intensity of the original light beam to increase or decrease, depending on how near the object is to the beam. This characteristic of electromagnetic radiation is known as the Fresnel effect, and light and radio waves are subject to the same laws of physics.

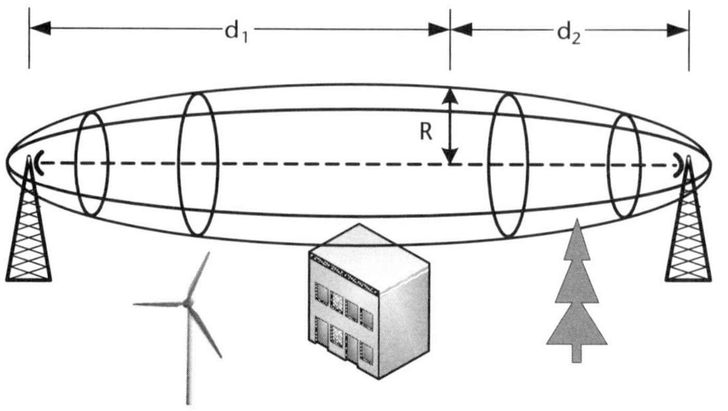

Figure 4.9 Radius of the first Fresnel zone.

Diffraction theory indicates that the direct path between the transmitter and the receiver needs a clearance of at least 60 percent of the radius of the first Fresnel zone to achieve required free-space propagation conditions. The higher the frequency, the narrower the Fresnel zone and, consequently, the more vulnerability to non-LoS effects (object attenuation) the radio link will experience.

The general formula (assuming that $R_n << d_1$ and $R_n << d_2$) to calculate the radius of the *n*th Fresnel zone at a certain location is approximated by:

$$R_n = 72.1\sqrt{\frac{nd_1d_2}{f(d_1+d_2)}} = \sqrt{n}\cdot R_1 \tag{4.12}$$

From here, for the first Fresnel zone ($n = 1$), we have:

$$R_n = 72.1\sqrt{\frac{d_1d_2}{f(d_1+d_2)}}\ \text{[feet]} \tag{4.13}$$

where

d: distance between antennas in miles ($d = d_1 + d_2$);

R: first Fresnel zone radius in feet;

f: frequency in gigahertz.

There is a critical difference between optical LoS (also known as *visual LoS*) and *radio LoS* (or radio visibility). Visual LoS considers only optical visibility (as seen by the human eye or aided by binoculars) between the ends of the path. Radio LoS takes into account the concept of Fresnel ellipsoids and their clearance criteria, and under normal atmospheric conditions, the radio horizon is around 30 percent beyond the optical horizon.

The Fresnel zone is computed along the path, usually for the distance of each of the terrain points, so the resolution of the computed and plotted Fresnel zone is comparable to the terrain data (Figure 4.10).

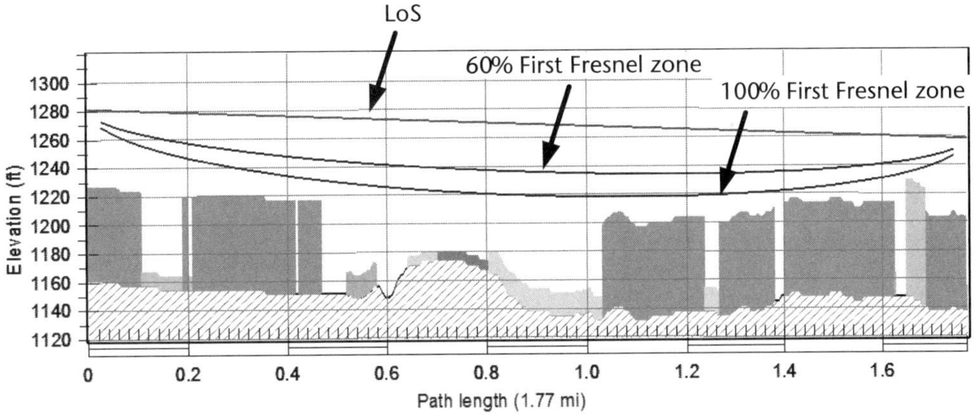

Figure 4.10 Microwave link path profile.

Typically, the first Fresnel zone is used to determine obstruction loss, and anytime the path clearance between the terrain and the LoS path is less than $0.6F_1$ (60 percent of the first Fresnel zone radius), some *diffraction loss* will occur and the amount of loss depends on the amount of penetration. Profiles are often drawn with the first Fresnel zone and a ratio of 0.6 to provide a quick visual inspection of possible problems caused by obstructions penetrating that zone.

The refractive properties of the atmosphere are not constant, and the variations of the *index of refraction* in the atmosphere (expressed by the Earth-radius factor k) may force terrain irregularities to totally or partially intercept the Fresnel zone.

4.4.4 Radio Link Budget

Engineers start the radio link design with a link budget analysis. The link budget is a calculation involving the gain and loss factors associated with the antennas, transmitters, receivers, transmission lines, antennas, and propagation environment used to determine the maximum distance at which a transmitter and receiver can successfully operate. (See Figure 4.11.) Antennas shown are parabolic dishes, but they could be Yagi, log-periodic, or any other type.

The purpose of *the transmitter* is to generate the carrier frequency that is to be used for the communication, to modulate this carrier frequency with the desired information, and finally, to amplify the signal so that it attains a sufficiently high power level to travel the desired communication distance to the receiver.

The *receiver* amplifies the received signal (which is at this point much weaker than when it was transmitted), filters out any undesirable signals (interfering signals) that the receiver picked up, and finally, detects the existence of information in the carrier frequency.

The purpose of the *feeder cable* is to interconnect the antenna with the transmitter/receiver. Feeder cabling between the radio equipment and the antenna may consist of coaxial cabling or a (flexible) waveguide.

The *antenna-coupling unit* makes it possible to utilize a common antenna for both the transmitter and receiver. The transmitter and receiver can, for example, be connected to the same antenna either via a duplex filter or a transmitter/receiver

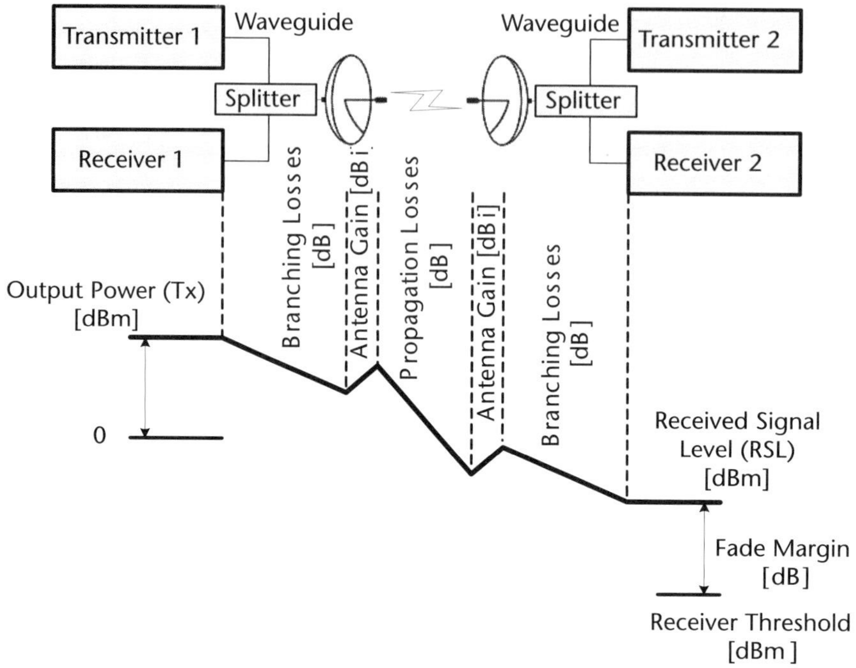

Figure 4.11 Radio path link budget.

switch. Together, feeder cable losses, antenna-coupling losses, and any additional losses (depending on the radio configuration) constitute *branching losses*.

The *antenna* adapts the generated signal to the surrounding environment (to the propagation medium) and directs the radio waves that are to be transmitted toward the receiving station. When receiving, the antenna receives the signal from the desired direction and sends it to the receiver. Every antenna is typically characterized by its impedance, bandwidth, directivity (radiation pattern), and polarization.

The *receiver sensitivity threshold* is the signal level at which the radio runs continuous errors at a specified bit ratio.

System gain (in decibels) is defined as the difference between the transmitter output power and the receiver threshold. Lowering the system gain will reduce the fade margin. High system gain can be used to reduce antenna sizes or improve the path availability. A given radio system has a system gain that depends on the design of the radio and the modulation used.

The gains from the antenna at each end are added to this gain. The free-space loss of the radio signal as it travels over the air is then subtracted from the system, and the longer the link, the higher the loss. These calculations result in a *fade margin* for the link, which is a difference between the received signal and receiver threshold value (or sensitivity) for a given BER. The radio can handle anything that affects the radio signal within the fade margin. If the margin is exceeded, then the radio link could become unavailable.

In most applications, the same duplex radio setup is applied to both stations forming the radio link. Thus, the calculation of the received signal level is independent of direction.

The *receiver signal level* (RSL) can be calculated as follows:

$$RSL = P_o - L_{ctx} + G_{atx} - L_{crx} + G_{arx} - FSL - L_m \quad (4.14)$$

where

P_o = output power of the transmitter (dBm);

R_x = receiver sensitivity threshold (dBm);

L_{ctx} = loss (cables, connectors, branching unit) between transmitter and antenna (dB);

L_{crx} = loss (cables, connectors, branching unit) between receiver and antenna (dB);

G_{atx} = gain of transmitter antenna (dBi);

G_{arx}= gain of receiver antenna (dBi);

FSL = free-space loss (dB);

L_m = miscellaneous losses (e.g., gas attenuation, obstacle losses, and misalignment losses expressed in dB).

The *effective isotropic radiated power* (EIRP) is defined as the effective power found in the main lobe of a transmitter antenna relative to an isotropic radiator, which has 0 dB of gain. It is equal to the sum of the antenna gain (in decibels relative to an isotropic radiator) plus the power (in decibels per milliwatt) into that antenna.

For example, if a 30-dBi gain antenna is fed with 27 dBm of power (including all the losses), it has an EIRP of 30 dBi + 27 dBm = 57 dBm.

4.4.5 Effects of Multipath, Rain, Dust and Sand

For a well-designed path that is not subject to diffraction fading or surface reflections, multipath propagation is the dominant factor in signal fading below about 10 GHz. The necessary reduction in path length and increase in frequency reduces the severity of multipath fading.

Rain fade refers primarily to the absorption of a microwave RF signal by the atmospheric rain, snow or ice, and losses that are especially prevalent at frequencies above 10 GHz. Rain fade is not limited to terrestrial point-to-point microwave links; it also can affect satellite uplinks or downlinks.

Above 10 GHz, the effects of precipitation tend increasingly to determine the permissible path length through the system availability objectives. The multipath and rain are two principal causes of fading, and they are normally mutually exclusive.

Given the split between availability and error performance objectives, precipitation effects contribute mainly to unavailability and multipath propagation mainly due to error performance. Another influence of precipitation (e.g., backscatter from rain) may influence the choice of RF channel arrangements.

Two countermeasures to propagation impairments are commonly used: *diversity techniques* (space, frequency, and hybrid) and *adaptive channel equalizers*, which attempt to combat attenuation and distortion caused by the propagation medium. *Adaptive modulation* is a new approach to combat rain problems in microwave point-to-point links.

Propagation through dust and/or sand is less researched and understood than multipath, and propagation through rain or even ducting. The atmosphere contains a variety of solid particles in suspension; in soil mechanics, particles with radii less than 60 microns are termed *dust*, depending on their size [15]. Particles with radii greater than 60 microns are typically called *sand*. The content of dust and sand will be vastly different in different parts of the world. Dust particles are random in shape and cannot be classified as spheres, ellipsoids or otherwise, which complicates their mathematical modeling.

Research has identified eight distinct dust storm types: planetary winds, cyclogenic, frontal, katabatic winds, haboob[22], constriction, dust devils, and diurnal winds [16]. The eight dust storm types were classified into a range of average wind speeds, maximum gusts, average storm widths, average storm lengths, and effective storm heights [17].

Data measured for an 11-GHz microwave link during a dust storm indicated a median signal reduction of 2–4 dB, while signal fades as high as 10–15 dB have been observed. This discrepancy of theoretical calculated and actual results can be explained by anomalies within and surrounding the dust storm, which practically produces its own microclimate and radio propagation conditions not trivial to model mathematically. [18]

After calculating link budget as shown in the previous section, the next step is to analyze rain fading, multipath fading, interference, and other (miscellaneous) losses that may affect the radio signal. This analysis is outside of the scope of this book but more details can be found in the literature.

4.5 Radio Systems in Electric Utility Space

All utilities—electric, gas, water and wastewater—deliver vital services to customers, and customers expect these services to be operational at all times. Key challenges shared by utilities include reliability, fast response to problems, and efficiency. Many utilities have installed smart grid systems—including two-way communications—to improve efficiency, security, reliability, and resiliency.

When implementing a smart grid, utilities install intelligent devices in their distribution system, management and control software in substations and their data center, and two-way smart grid communications to link the devices and software. A smart grid enables new applications such as AMI, DA, outage management, automatic load shedding, and the ability to manage alternative energy sources.

In addition, the same wireless smart grid communication network that supports these applications can also enable mobile workforce applications in the field.

22. The name comes from the Arabic word habb, meaning "wind." Haboob is a strong sand and dust storm that occurs along the southern edges of the Sahara, the Sudan. In some areas of North America, sand- and dust storm walls associated with severe thunderstorms are also called haboobs. They are especially frequent in Arizona and Texas with wind speeds around 30–50 mph.

4.5.1 Microwave Point-to-Point Systems for Utilities

For the reason of increased reliability it is common to use licensed microwave radios for long-distance and/or high-bandwidth applications, as well as for critical applications within electric utilities. Licensed frequency bands are coordinated by regulating bodies to ensure interference-free operation.

License-exempt frequency bands are not coordinated and, therefore, are not given any legal recourse by regulating bodies in the event of interference.[23] Therefore, microwave systems using license-exempt bands should not be used to transport electrical utility critical communications traffic (for protection, for example), except in cases of secondary (redundant) communications paths or maybe as a temporary solution. License-exempt radio devices on the *smart grid* operate in one of the bands set aside by the FCC—for the most part either 902–928-MHz or 2.4-GHz bands, and to some degree at 5.8 GHz.

Availability and reliability of license-exempt microwave systems is not considered sufficient for the highly important communications system on whose operation and performance the electrical grid depends.

Registered frequency bands (sometimes also referred to as *lightly licensed frequency bands*) are similar to license-exempt frequency bands in that there is no recourse in the event of interference. The advantage of the registered band is the registration requirement that allows users to coordinate among themselves and mitigate any interference issues that may arise. Throughout most of the world a band centered at 3.5 GHz is available for public access data networks and is fairly widely used. Early in 2005, the FCC approved new spectrum for broadband data services located between 3.650 and 3.700 GHz (WiMAX).[24]

In addition, many new radios have a built-in *encryption* as an option that can be enabled if required by the user. Figure 4.12 depicts a typical all-indoor microwave system commonly used by power utilities.

In 2003, the FCC authorized a 50-MHz radio frequency spectrum from 4,940 to 4,990 MHz for noncommercial use by state and local governments and public safety organizations that are dedicated to the protection of life, health, or property. This way, the FCC ensured a space in the airways free from interference by corporate or general public communications.[25] This band is very sparsely used by public safety organizations, and power utilities are trying to get access to it for their use.

Microwave systems from 3 GHz to approximately 10 GHz are relatively limited in throughput, do not readily conduce to high degrees of frequency reuse, and, perhaps most important, share a vulnerability to what is known as *multipath distortion*. Frequencies between 10 and 40 GHz are widely used by point-to-point microwave communications systems worldwide, especially after the invention of split-mount microwave radios, but they all share susceptibility to rain and have to be carefully designed to provide sufficient fade margin to combat effects of rain.

23. Not all the license-exempt radio systems utilize spread-spectrum or frequency-hopping techniques.

24. FCC is considering important changes to this band. FCC adopted three-tier licensing system in its ruling FCC-15-47 AI on April 17, 2015.

25. Speaking of public communications, there are some areas of U.S. in which certain types of radio operation are not allowed. For specific locations, consult the FCC Rules Part 2, Section 106, under Footnote US311 for Radio Astronomy Quiet Zones.

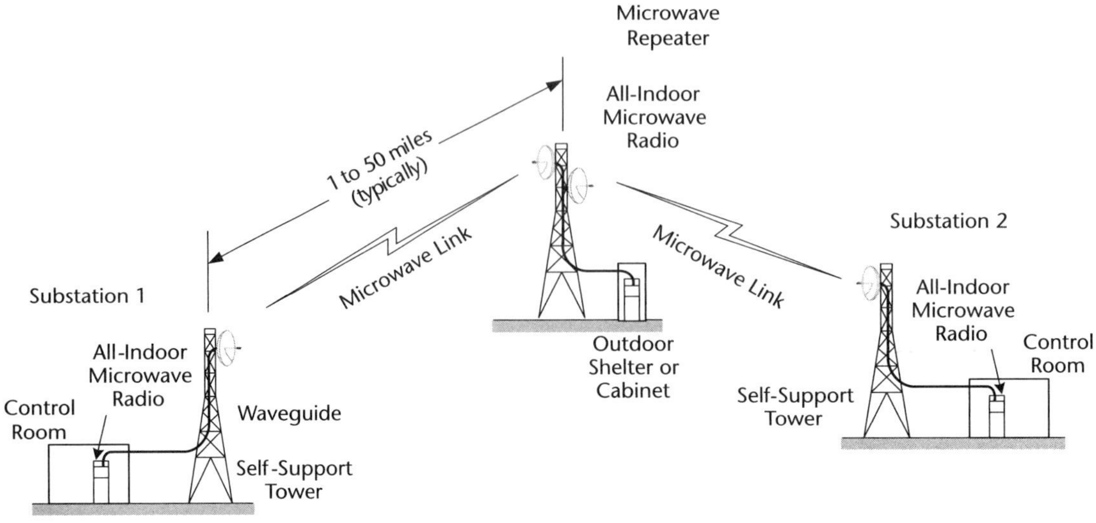

Figure 4.12 Microwave point-to-point system.

In the United States, although power utilities use different bands, it seems that 6- and 11-GHz are the two most popular licensed microwave frequency bands and that 900 MHz and 5.8-GHz are the most popular license-exempt frequency bands. There are no frequency bands dedicated strictly to power utility applications; they have to share frequency bands with other users and applications.

4.5.2 AMI

AMI allows utilities to collect, measure, and analyze energy consumption data for grid management, outage notification, and billing purposes via two-way communications. While a predecessor technology called automatic meter reading (AMR), still in use today, uses one-way communications to accomplish meter readings primarily for monthly billing purposes, AMI can be leveraged to provide consumers with historical energy consumption data, comparisons of energy use in similar households, dynamic pricing information, and suggested approaches to reducing peak load via in-home displays [19].

An AMI system consists of a number of technologies and applications that have been integrated to perform as one—smart meters, wide-area communications infrastructure, HANs, meter data management systems (MDMSs), and operational gateways.

Determining the appropriate communications technologies for AMI applications will depend on the level of AMI functionality desired. Early AMI installations have traditionally been serviced by PLC technology, which is used for relaying meter data and other internal communications over a utility's power lines. PLC is still a very common technology for AMI functions in rural, low-density areas.

Fixed point-to-multipoint radio networks, also known as star, radial, or hub-and-spoke networks, are common in current installations, using licensed spectrum.

Many AMI deployments, particularly in urban areas, use 900-MHz licensed or license-exempt radio mesh networks. Mesh networks for AMI, commonly referred to as NANs, benefit from the resiliency of mesh architecture with its multiple radio paths to each endpoint. In a mesh network, each endpoint has the ability to function as a router, and connectivity between meters and collection points is typically achieved via a dedicated network using license-exempt radio spectrum, run either by the utility or a subcontractor [20].

Multiple encryption standards are used to protect customer and utility meter data (e.g., 128-bit AES encryption or 256 EC[26] encryption, and digital signatures).

4.5.3 DA

DA allows utilities to remotely monitor and control assets in its distribution network through automated decision-making, providing more effective fault detection and power restoration. DA includes control center–based control and monitoring systems, such as distribution SCADA or distribution management systems, and distribution automation field equipment, ranging from remote terminal units to intelligent electronic devices such as circuit breakers, reclosers, switches, capacitors and transformers that can be remotely monitored—and remotely controlled or operated.

DA represents one of the least latency-tolerant smart grid applications, with requirements for less than 1 second of latency for alarms and alert communications and sub-100 milliseconds for messaging between peer-to-peer nodes inside RF mesh configurations. The maximum latency for certain more latency-tolerant DA applications will not exceed two seconds.

Bandwidth requirements are in the range of 9.6 kbps—100 kbps, and the required level of reliability will be 99 to 99.999 percent. As these functions are only required when there is electricity available, backup power is not required.

Utilities use a variety of communications technologies to provide DA functionality; the most common backhaul communications media being used for DA is the public cellular network, followed by the license-exempt 900-MHz spread spectrum,[27] and the licensed *multiple address radio* (MAS).

MAS systems are defined as fixed-radio, low-capacity, point-to-point or point-to-multipoint systems where a master transmit station will communicate with one or more remote facilities. The master stations are typically used to collect data from the remote station(s) and to send control commands to these remote station(s).[28] The master station connects to a data acquisition controller either directly or via

26. Elliptic curve cryptography (ECC) is a public-key cryptography based on the algebraic structure of elliptic curves over finite fields. Elliptic curves are applicable for encryption, digital signatures, pseudorandom generators and other tasks.

27. Most electric utilities would prefer to use licensed radio frequencies for DA, but there is a lack of spectrum availability in the 150–750 MHz range.

28. Typical MAS applications include utility company data collection; controlling railroad crossing signals; monitoring and controlling valves along gas and water pipelines; reclosers; and monitoring building security systems. There is a total of 3.2 MHz of spectrum in the 900 MHz band available for MAS usage. The specific frequency assignments are listed in Section 101.147 of the FCC Rules. Most MAS facilities are licensed under CFR 47, Part 101. The licensing form is the FCC Form 601.

dedicated wireline or wireless link. Each remote radio station connects to a data collection unit like RTU, *programmable logic controller*, or IED.

The master unit is always ready to transmit and receive to prevent delays due to transmitter keying. Each remote unit with a unique address is always in the listening mode until polled and is than ready to transmit. For a large service, area multiple master units are needed to cover groups of remote units. Each link covers typically 10 miles (16 km), which can be extended by using repeaters. The data rate is up to 4.8 kbps and can be increased to 9.6 kbps but the coverage reduces. The FCC allows channel bandwidth of only 12.5 kHz or less (6.25 kHz may be mandated in the future), which limits the expansion and upgrade capability of the MAS.

Units are typically polled in a round-robin fashion, although some work has been done to demonstrate the use of MAS radios in a contention-based network to support asynchronous remote device transmissions.

Some of the other technology and frequency alternatives available to utilities are briefly described in Section 4.5.4.

4.5.4 Radio Technology Options

Radio technology is becoming increasingly useful for accessing individual users and narrow- and medium-bandwidth groups of users, thereby replacing wireline or fitting into the bandwidth gap between wireline and fiber. Aside from point-to-point microwave systems, more applications are expected for WLANs, Bluetooth, ZigBee, and other new short-range access technologies in utility space (see Figure 4.13).

There is a new frequency band called *white space* that refers to frequencies allocated to a broadcasting service but not used locally that allow the use of license-exempt devices that can guarantee that they will not interfere with assigned

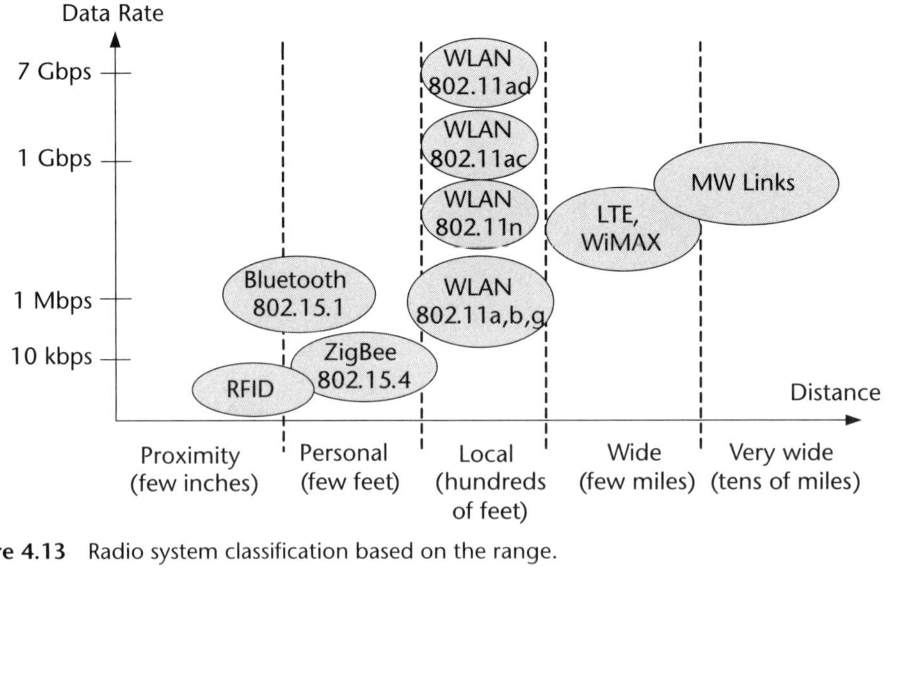

Figure 4.13 Radio system classification based on the range.

broadcasts.[29] The idea is that the spectrum could be used to augment existing radio services or eventually be used to create new wireless broadband services. [21]

In 2010, the FCC allowed license-exempt fixed and portable access over white spaces [22] and issued an updated ruling in 2012 [23]. The formal standards are IEEE 802.11af-2012 (Draft 2.0), which is still in draft version, and a published standard IEEE 802.22-2011 Wireless Regional Area Networks (WRAN).

Although conceptually interesting, it remains to be seen whether some meaningful applications for utilities will develop in white-space frequency bands.

4.5.4.1 WLANs

A WLAN is a wireless computer network that links two or more devices using radio transmission (often spread-spectrum or OFDM radio) within a limited area such as a home, school, office building, or substation. (See Figure 4.14.) Most modern WLANs are based on IEEE 802.11 standards.

Residential/business users of WLAN sometimes have difficulties related to factors such as signal strength and security, and, at first glance, power utilities are concerned about applying WLAN for industrial use. However, the industrial-grade WLAN technologies have already been in use for other mission-critical industrial applications, such as mobile traffic control and road signs in transportation industries; mining automation; video surveillance in financial industries; tracking inventories and assets; and pipeline monitoring.

Industrial-grade WLANs are hardened for a harsh environment (with qualities such as stronger signal, rugged outdoor design, redundant power supplies, and

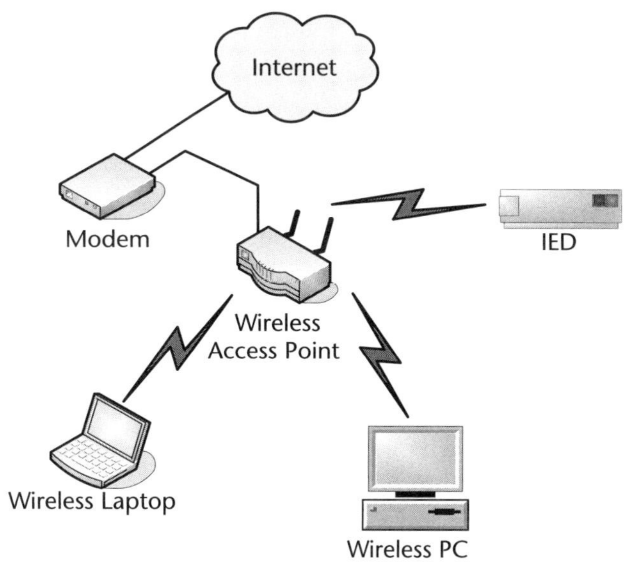

Figure 4.14 Basic components of a residential/business WLAN system.

29. White space exists in the broadcast TV operating frequencies known as the VHF/UHF band; it ranges from 470–790 MHz in Europe and non-continuous 54–698 MHz in the United States.

increased security), but they are also more expensive [24]. A brief description of available 802.11 standards are listed below:

- IEEE 802.11b: Operates in the license-exempt portion of the 2.4-GHz radio frequency spectrum; delivers a maximum data rate of 11 Mbps; and enjoys broad user acceptance and vendor support [25].
- IEEE 802.11a: Operates in the 5-GHz radio frequency spectrum. With a maximum data rate of 54 Mbps, this standard offers a fivefold performance increase over the 802.11b standard. It uses eight nonoverlapping frequency channels, resulting in increased network capacity, improved scalability, and the ability to create microcellular deployments without interference from adjacent cells. Operating in the license-exempt portion of the 5-GHz radio band, 802.11a is also immune to interference from devices that operate in the 2.4-GHz band, such as microwave ovens, cordless phones, and Bluetooth.
- IEEE 802.11g: A high-performance standard delivering the same 54-Mbps maximum data rate as 802.11a, but operating in the same 2.4-GHz band as 802.11b.
- IEEE 802.11n: An amendment to the IEEE 802.11-2007 radio networking standard, significantly improving network throughput over previous standards, such as 802.11b and 802.11g, with a significant increase in the maximum data rate from 54 Mbps to a maximum of 600 Mbps. This is achieved by adding multiple-input multiple-output (MIMO) and channel-bonding/40-MHz operation (two adjacent 20-MHz channels, bonded together) to the physical (PHY) layer, and frame aggregation to the MAC layer. The new standard, published in 2012, is referred to as IEEE 802.11-2012.
- IEEE 802.11ac-2012: A standard that is an extension of 802.11n and that will provide increased throughput in the 5-GHz band. This specification enables higher multistation WLAN throughput of at least 1 Gbps and a maximum single link throughput of at least 500 Mbps, by using wider RF bandwidth, more streams (up to eight), and high-density modulation, up to 256 QAM. All devices will have to support 20-, 40-, and 80-MHz channels. In addition, it provides optional support for operation on 160-MHz channels. Both 80- and 160-MHz channels can be formed by a combination of two adjacent nonoverlapping 40- and 80-MHz channels, respectively. Beamforming, and therefore reduced interference, is an important additional feature of the new standard (also available in 802.11n). The new feature of 802.11ac is multiple user MIMO (MU-MIMO), which allows stations having multiple antennas to transmit several data streams to multiple users at the same time over the same frequency channel.
- IEEE 802.11ad-2012: Also called "WiGig," a published and completely new standard that is already seeing a major push from hardware manufacturers. Extreme data capacity requires a large spectrum allocation, available only in very high-frequency bands. Using the 60-GHz license-exempt band, the new standard can achieve a theoretical maximum throughput of up to 7 Gbps, but at very short distances.

4.5.4.2 IEEE 802.15.1 (Bluetooth)

Bluetooth[30] radio technology built into equipment replaces the cable connecting the two devices. Bluetooth is a de facto open standard for short-range digital radio, and although it is considered to be a point-to-multipoint system, it can also be a point-to-point system, depending on the application.

Bluetooth operates in the 2.4-GHz ISM band, a license-exempt portion of the radio spectrum that is already well used and generally available in most parts of the world. It shares the band with microwave ovens and other RF communications technologies, most notable of which are HomeRF and IEEE 802.11b (WLAN) [26].

Bluetooth uses FHSS and is a shorter-range and lower-bandwidth technology than 802.11b, and it uses frequently changing, narrow bands over all channels[31]. It supports the bit rate of 1 Mbps or, with enhanced data rate, a gross air bit rate of 2 Mbps *(basic rate)* or 3 Mbps *(enhanced data rate)*.

Bluetooth 4.0. In April 2009, the Bluetooth SIG (Special Interest Group) formally adopted Bluetooth Core Specification Version 3.0 High Speed (HS), or Bluetooth 3.0. [27] The inclusion of the 802.11 *Protocol Adaptation Layer (PAL)* provides increased throughput of data transfers at the approximate rate of 24 Mbps. In addition, mobile devices including Bluetooth 3.0 will realize increased power savings due to built-in enhanced power control.

Adopted in 2010, Bluetooth 4.0 includes *classic Bluetooth*, *Bluetooth high-speed* and *low energy Bluetooth* (BLE) protocols. Bluetooth high speed is based on Wi-Fi, and classic Bluetooth and consists of legacy Bluetooth protocols. BLE is based on a different technology, previously known as WiBree, and it represents a subset of Bluetooth 4.0 with an entirely new protocol stack for rapid buildup of simple links. As an alternative to the Bluetooth standard protocols that were introduced earlier, it is aimed at very low-power applications [28].

For BLE, data throughput is not a meaningful parameter since it does not support streaming. Although it has a data rate of 1 Mbps, it is not optimized for file transfer, but for sending small chunks of data at the time and exposing state. In other words, it is good at small, discrete data transfers, triggered by local events, and as such, BLE is sometimes seen as a competition and possible replacement for ZigBee. BLE is a point-to-multipoint (star) PAN technology having a very short range.

A useful range in open space is 150m (500 ft), 10 mW power (10 dBm), GFSK[32] modulation, 3 ms latency, adaptive frequency hopping, 128 AES, and star topology.

---

30. The Bluetooth standard is named after Harald I Bluetooth, the king of Denmark between 940 and 985 AD who united Denmark and Norway. Bluetooth technology unites devices via radio connections, hence the inspiration for its name. The Bluetooth standard was developed jointly by 3Com, Ericsson, Intel, IBM, Lucent, Microsoft, Motorola, Nokia, and Toshiba.

31. It is important to manage the concurrent operation of 802.11b WLANs and Bluetooth within the enterprise. Task Group 2 of the IEEE 802.15 Working Group is looking at the coexistence issues of IEEE 802.11b WLANs and Bluetooth.

32. Frequency shift keying (FSK) conveys information by decreasing the carrier frequency for the duration of a 0 symbol and increasing the frequency for the duration of a 1 symbol. In GFSK, a Gaussian filtering (called "pulse shaping") is applied to the square-wave signal that would shift the carrier frequency and thus reduce the spectral bandwidth and out-of-band spectrum, increasing adjacent-channel power rejection.

4.5.4.3 IEEE 802.15.4 (ZigBee)

ZigBee takes its name from the dance that honey bees use to communicate information about newly found food sources to other members of the colony. WSN nodes use ZigBee to communicate in a mash topology. In mesh networks, each wireless node communicates with the one adjacent to it, and in the event of node failure, information gets automatically rerouted to allow devices to continue communicating.

ZigBee Alliance[33] is testing a new unified wireless standard called *ZigBee 3.0*. Based on IEEE 802.15.4, which operates at 2.4 GHz, it uses ZigBee PRO networking to enable reliable communication in the smallest, lowest-power devices.

ZigBee Smart Energy is the standard for interoperable products that monitor, control, inform, and automate the delivery and use of energy and water. It is used to deliver innovative solutions for smart meters and the HAN that allow consumers to know and control their energy use by connecting them to the smart grid and helps create greener homes by giving consumers the information and automation needed to easily reduce their consumption and save money.

ZigBee can be utilized in NANs, which are designed to provide utilities last-mile connectivity for smart meters and distribution automation devices to WAN gateways, such as RF collectors or data concentrators and field devices. Research has shown no significant adverse impact on the performance of ZigBee by the electromagnetic environment of the substation, and therefore, it can be also used for automation purposes inside a substation [29].

4.5.4.4 WiMAX

WiMAX is a technology based on the IEEE 802.16 air interface standard suite, which provides the radio technology for nomadic and mobile data access. The first version of WiMAX was approved as the 2002 IEEE Standard 802.16-2001 that addresses fixed LoS connections by focusing on licensed frequencies in the range of 10–66 GHz.

WiMAX offers an alternative to cabled networks such as fiber-optic links, traditional cable or digital subscriber lines, or T1/E1 networks [30]. The *IEEE 802.16a-2004* standard (Revision of IEEE Std 802.16-2001) is designed for stationary transmission, and the *802.16e-2005* amendment deals with both stationary and mobile transmissions. The latest version added coverage for the 2–11-GHz range, enabling support for both licensed and licensed-exempt bands, as well as non-LoS operation.

WiMAX employs OFDM[34] and supports *adaptive modulation and coding* depending on the channel conditions. Radio systems cover large geographic areas with the maximum data rate for 802.16a around 75 Mbps, and the maximum range 30 miles but, obviously, not at the same time. The maximum data rate for 802.16e is 15 Mbps and the range 1–3 miles. There is no uniform global licensed

33. The ZigBee Alliance, which totals nearly 100 companies, including Honeywell, Mitsubishi, Motorola, Philips, and Samsung, is hoping to penetrate a home-automation market confused by a variety of other proprietary technologies.

34. The basic idea behind OFDM is the use of a large number of parallel narrowband sub-carriers instead of a single wideband carrier to transport information. OFDM technology has been known for decades but its complexity and high cost have prevented widespread use.

spectrum for WiMAX, although the WiMAX Forum has published three licensed spectrum profiles (i.e., 2.3 GHz, 2.5 GHz, and 3.5 GHz).

A relatively low-cost, high-performance solution, WiMAX is suitable for most noncritical smart grid applications, including substation automation, DA, video monitoring, and mobile workforce applications. WiMAX can be deployed cost-effectively in a private network by utilities concerned about reliance on public networks.

WiGRID[35] is a new *WiMAX Forum*[36] technology based on IEEE 802.16e, intended for utilities and industrial applications. It is an Ethernet technology designed for wide-area networking, primarily for layer 2 networking, and optimized for fixed applications, but if needed, it can also support nomadic and mobility applications.

4.5.4.5 Long Time Evolution (LTE)

LTE is a mobile technology upgrade project for 3G networks initiated in 2004 to provide faster data speeds and new services through new radio access technology optimized for IP-based traffic. Performance targets included average downlink user throughput of 100 Mbps and 50 Mbps average uplink throughput. LTE is extremely flexible, using a number of defined channel bandwidths between 1.25 and 20 MHz.

LTE downlinks use *OFDM access* (OFDMA) to achieve high peak data rates in high-spectrum bandwidth and support data-modulation schemes QPSK, 16QAM, and 64QAM. LTE uplinks use *single-carrier-frequency division multiple access (SC-FDMA)* and support BPSK, QPSK, 8PSK, and 16QAM modulation, while E-UTRA defines the radio interface.

To suit as many frequency band allocation arrangements as possible, both paired (FDD) and unpaired (TDD) band operation is supported. LTE can coexist with earlier 3GPP radio technologies, even in adjacent channels, and calls can be handed over to and from all 3GPP's previous radio access technologies. [31].

LTE is a high-speed, high-capacity radio communications standard with low latency, as well as flexibility in prioritization and in QoS that will enable its use in critical applications, including utility space. LTE may be used for substation automation, DA, video monitoring, and mobile workforce management applications in the smart grid, and the standard is considered future-proof by many utilities. In other words, LTE is well-suited to *field area network* (FAN) communication to distributed devices (i.e., "last-mile" connectivity in the utility environment).

The core concept is that a utility can use a commercial cellular network to connect its devices. In theory, this lets the utility focus on its core business and turn over the communications issues to a specialist. Some of the advantages are that there is no last-mile connection build-out needed since the existing infrastructure

35. WiGRID is a new IEEE 802.16 system profile being developed by the WiMAX Forum for the requirements of the utility companies. WiMAX Forum formed the "Smart Grid Working Group" in March 2011 and first meeting was held in May 2011.

36. The WiMAX Forum is an industry-led, not-for-profit organization that certifies and promotes the compatibility and interoperability of broadband wireless products based on IEEE Standard 802.16. The IEEE 802.16 standard was developed to deliver non-line-of-sight (NLoS) connectivity between a subscriber station and base station.

is used and the technology is proven and matured. Some type of partnerships with the federal government or with incumbent LECs may allow U.S. utilities to share the costs for LTE network deployment and operation.

There are some disadvantages as well; change of the technology, spectrum availability, price, insufficient reliability[37], and probably most important of all for power utilities, loss of control; as a result, many utilities, particularly in the United States, remain apprehensive of relying on public communications providers.

4.5.4.6 RF Identification (RFID)

RFID was introduced in October 1948 in the paper of Harry Stockman titled "Communications by Means of Reflected Power." The popular system identification of friend of foe (IFF) for aircraft was one of the first applications of the RFID [32]. In the early 1940s, the British Royal Air Force outfitted airplanes with radio transponders that would respond when interrogated. This allowed pilots and ground crews to distinguish the RAF airplanes from the Luftwaffe's, which proved to be a decisive advantage in the battle of Britain.

In 1960s, the electromagnetic theory related to the RFID application was developed, and this was the prelude to RFID expansion. Commercial activities exploiting the RFID began also during the 1960s; the *electronic article surveillance* (EAS) application is one example.[38]

In general terms, RFID represents a way of identifying objects or people using radio waves. Identification is possible by means of unique numbers that identify objects, people, and information, stored on microchips, which can be read automatically, unlike bar codes that need to be scanned manually [33]. RFID systems are comprised of three main components, described as follows:

- The RFID tag, or transponder, which is located on the object to be identified and is the data carrier in the RFID system;
- The RFID reader, or transceiver, which may be able to both read data from and write data to a transponder;
- The data processing subsystem, which utilizes, in some useful manner, the data obtained from the transceiver.

The essential requirement in an RFID system is to transfer data stored in a tag, to a reader, across a radio air interface (the region between the tag and the reader). A two-way communications process is required to do this and necessitates a radio carrier signal suitably modified (modulated) to carry the data. The concept is depicted in Figure 4.15.

37. One of the common problems during widespread disasters is that the radio cell sites can be damaged, no power or exhausted battery backup, or simply be overwhelmed by the nonemergency traffic. Although many cell sites (including those with large tower structures) have battery backup, the wireless industry has successfully been resisting FCC-mandated 24-hour backup requirement. Even among the sites that do have battery backup of two or more hours, due to the lack of maintenance, many of those batteries may not be operational.

38. The EAS is a simple one-bit tag, since only the presence or the absence of a tag can be detected. Many items in stores are equipped with this type of tag to prevent theft.

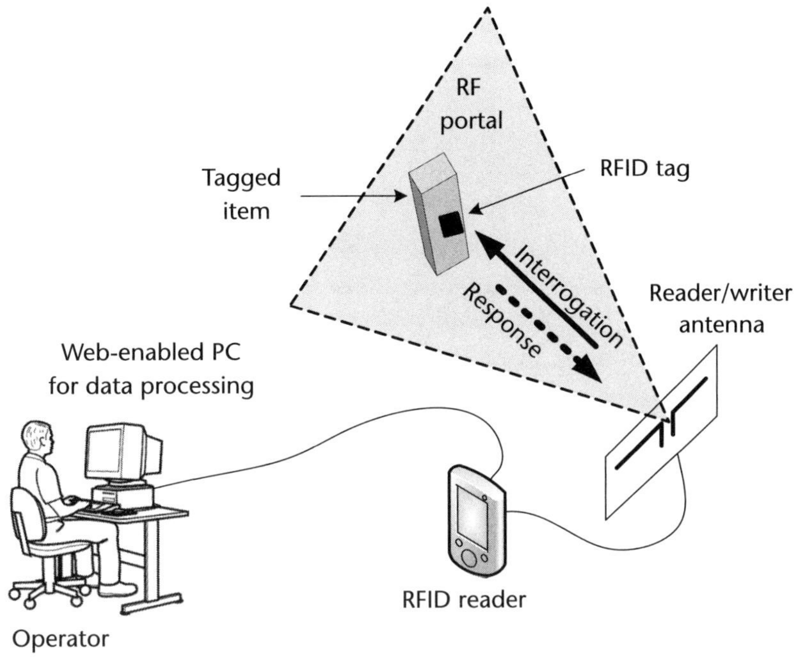

Figure 4.15 Basic components of RFID system.

A reader transmits a signal that is received by an antenna integrated with a small RF chip that is activated only when an RFID reader scans it. When the chip wakes up, it sends the unique identifier number, which the reader passes along to applications, such as inventory control and shipping. It is this level of application logic that provides the selection of particular tag and data manipulation criteria for those identified tag(s). Using a host computer application program, specific tags can be selected to be identified, and the data contained within those tags to be acted upon while those tags are active within the *RF portal* (interrogation area). The RF portal is defined as the area where RFID tags can be read or written to; it can be stationary or mobile.

RFID operates in license-exempt spectrum space, but the exact frequencies that constitute license-exempt bands may vary depending on the regulations in different countries. Typical carrier frequencies (reader's transmitting frequency) in today's applications range from 125 kHz to 2.45 GHz (with 5.8 GHz also being considered). The frequency bands must be selected carefully for applications because each one has its own advantages and disadvantages.

Two of the most predominant RFID applications have been in the form of toll road collection transponders and security badges. Toll road authorities around the country have equipped drivers with a transponder that is connected to their credit card. Security badges have been equipped with RFID chips to allow centralized control of access to facilities (like a substation, for example) and specific rooms within buildings, switch yards, control rooms, and the like. These can also be used to track the locations of people in a facility by identifying the door they last passed through.

4.5.5 Radio Network Deployment

4.5.5.1 Designing Privately Owned Radio Networks

Radio communication often plays a significant role in the modernization of utility networks, with a number of different radio technologies and architectures used by different organizations today (Figure 4.16).

Every radio network is unique, and the build-out process requires a number of well-planned and properly executed steps.

From the topology perspective, SG is utilizing a combination of different approaches—from point-to-point, to point-to-multipoint, and mesh networking. These approaches are described as follows.

- Point-to-point communication systems provide connection between two locations. They offer full bandwidth while latency is below 0.15 milliseconds. But these systems usually require LoS between the antennas at the endpoints, so they can be difficult to deploy in hilly terrains.
- Point-to-multipoint communication systems connect multiple remote sites to a master site location. Bandwidth is shared between users at each site, and is typically limited to 50–100 Mbps per site, while latency can exceed 50 milliseconds. Since point-to-multipoint systems are susceptible to a single point of failure at the master site, they are not well-suited for SG applications other than AMI backhaul.
- Mesh communication systems eliminate single-point congestion and provide the highest possible reliability, as there is no single point of failure.[39] These systems typically offer latency of up to 100 milliseconds. Mesh systems continually evaluate signal quality and select the best path between points, keeping optimal performance and reliability. Since they provide the best overall performance, mesh systems are preferred for most SG applications.

Design requirements for utility radio networks will differ from the usual wireless carrier networks; therefore, design and deployment will utilize a different approach than carrier wireless networks. One of the main differences is the predefined (for most cases) existing sites that have to be covered with the radio signal of the sufficient strength. Therefore, there is no flexibility in finding the best location for the *base station* (BS) covering that area, which makes radio design more challenging [34].

Regardless of the type, application, frequency, or network size, there are six basic steps that should be executed to successfully build an outdoor radio network:

1. Radio network planning (including such steps as radio propagation feasibility study, researching network requirements, obtaining digital terrain data, examining regulatory issues and licensing, and issuing preliminary design plans).
2. Equipment selection (RFIs and RFQs).

39. The original concept of a mesh network was developed for military applications by DARPA (Defense Advanced Research Projects Agency) in the 1980s. It was designed for battlefield data communications between moving vehicles and used an architecture optimized for vehicle-to-vehicle (peer-to-peer) communications.

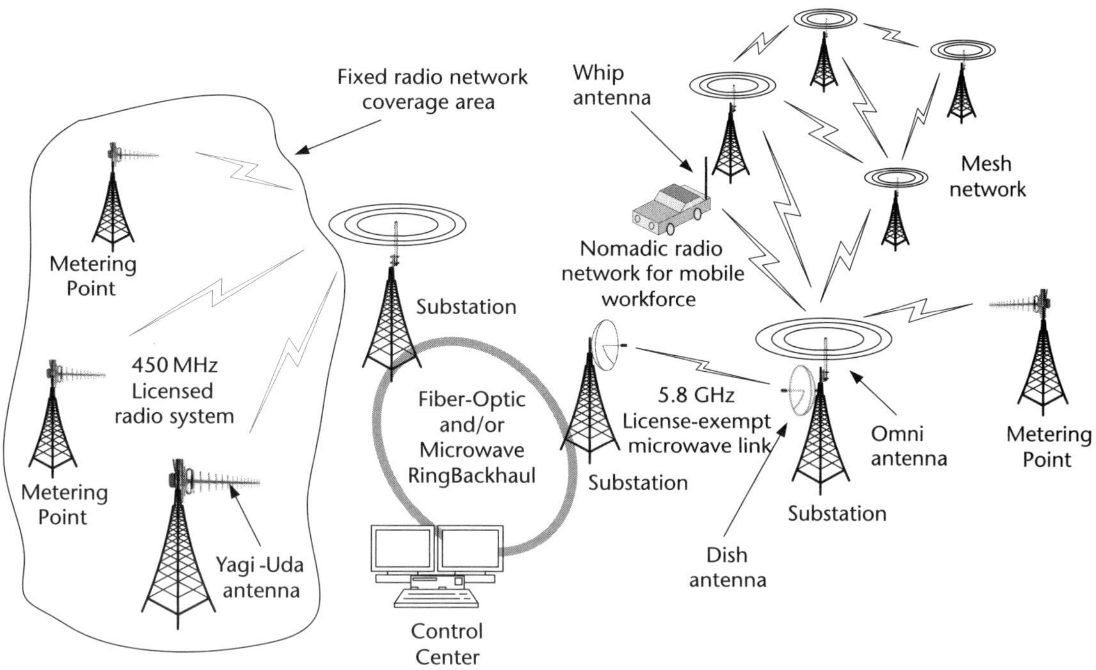

Figure 4.16 Radio network architecture.

3. Test sites and field surveys (e.g., installation, testing, correction of design models, and confirmation of clutter data).
4. Radio network design (e.g., topology selection, licensing, permitting, site and path surveys, and infrastructure selection).
5. Network deployment (e.g., contractor selection, construction, and equipment installation).
6. Testing and commissioning (including as-built documentation).

4.5.5.2 Radio Propagation Study

Radio network planning is an iterative process requiring multiple steps, and a vendor-independent radio network propagation study (feasibility study) is an essential first step in the process of designing reliable radio networks for SCADA, voice, video surveillance and Internet applications.

As mentioned earlier, clear LoS is a preferred propagation mechanism in every radio network. Some systems absolutely require clear LoS, while others can function without it.

There are several methods for estimating the radio propagation that will occur in the field. Many of the models that describe outdoor radio propagation and environment influences on the strength of the radio signal at a certain location are based on probability theory and field measurements, and thus semi-empirical in nature (e.g., Okumura/Hata, Longley-Rice, and ITU curves). On the other hand, some of the models are deterministic [i.e., the calculations are based on the actual terrain profile and clutter data (TIREM[40], National Spectrum Management Association, and Pathloss)].

The radio propagation study is a desktop study with the purpose of determining radio signal strength at the locations of interest. The study will point out potential degradation in radio propagation and enable the next steps toward detailed radio and/or microwave network design and equipment selection.

In rural environments, the number of sites to be connected into the network may be small, but the sites may be spread out over a very large geographic area, sometimes thousands of square miles, requiring a different approach—possibly a combination of different technologies.

Although radio propagation study for utility applications usually encompasses only a fixed-data network, the same analysis can be done separately for other types of radio network (fixed, nomadic, or both).

4.6 Review Questions and Problems

1. Verify that the correct value for the speed of light in vacuum, c, is obtained when numerical values for the permeability and permittivity of free space (μ_0 and ε_0) are entered into the equation:

$$c = \frac{1}{\sqrt{\mu_0 \varepsilon_0}}$$

2. There are many types of antenna radiation patterns, but the most common are omnidirectional, pencil beam, fan beam, and shaped beam. Discuss them in more detail and find examples of applications. More information can be found in [35].
3. Calculate the wavelength (in inches) of the radio signals that have frequency of:
   A. 100 kHz ($\lambda =$ ____ in)
   B. 915 MHz ($\lambda =$ ____ in)
   C. 5.8 GHz ($\lambda =$____ in)
4. Express 3th Fresnel zone radius as a function of the first Fresnel zone radius, R.
5. Compare the width of the Fresnel zone at its widest point, for 10-mile microwave links, operating at 6 and 11 GHz. From the frequency selection perspective, if there is a danger for a part of the Fresnel zone to be obstructed by the tree, which frequency would you prefer to use in your design? Write a detail explanation of your answer.
6. Verify whether the following statements are correct and justify the answer:

40. TIREM stands for Terrain Integrated Rough Earth Model. TIREM predicts median propagation loss from 1MHz to 40 GHz. As opposed to Longley-Rice, TIREM has built-in routines for evaluating radio paths over sea water. Because the TIREM is a proprietary model it is not possible to tell exactly what its code is doing, which makes the model less attractive to the FCC and other users.

A. A horizontally positioned, linearly polarized receiver antenna is unlikely to capture any of the energy emitted by a vertically positioned, linearly polarized transmitter antenna.

B. Two antennas can communicate even if they have different polarizations, assuming they are sufficiently close to each other.

C. The term "radio communications system" is more technically correct and should be used instead of "wireless communications system."

7. Are there any known and documented blackout fading areas? Research and list some of the areas of verified blackout fading occurrence.

8. What are the pros and cons of using microwave radios with higher modulation rate?

9. A 2-foot parabolic dish antenna is operating at 11 GHz. Determine the approximate antenna gain, beamwidth, and distance for the far-field region operation. Assume illumination (radiation) efficiency of 55 percent.

$$D = 2\text{ ft} = 24\text{ in}$$

$$\lambda = \frac{c}{f} \approx 0.0273\text{ m} = 0.0896\text{ ft}$$

$$G = 20\log 2 + 20\log 11 + 7.5 \approx 34.3\text{ dBi}$$

$$BW = \frac{70}{f \cdot D} = \frac{70}{11 \times 2} \approx 3.2^\circ$$

$$R_{FF} = \frac{2D^2}{\lambda} = \frac{2 \cdot 2^2}{0.0896} = 89\text{ ft to infinity}$$

A. Repeat the calculation for 6 GHz frequency.

10. MW link has the following parameters: frequency = 11 GHz, path length = 7.5 miles, antenna gain = 41 dBi, miscellaneous antenna/T_X/R_X losses = 0.5 dB, transmitter power = 0.13W, maximum receive power = –15 dBm, Rx threshold level = –72 dBm @10^{-6} BER. Calculate the EIRP, FSL, RSL, and FM for this microwave link.

$$EIRP = 21 + 41 - 0.5 = 61.50\text{ dBm}$$

$$FSL = 96.6 + 20\log 7.5 + 20\log 11 = 134.93\text{ dB}$$

$$RSL = 21 + 41 - 0.5 - 134.93 + 41 - 0.5 = -32.93\text{ dBm}$$

$$FM = -32.93 - (-72) = 39.07\text{ dB}$$

A. Repeat all the calculations for a 15-GHz, 6-mile link, keeping the rest of the parameters the same (Answer: EIRP = 61.50 dBm, FSL = 135.68 dB, RSL = –33.68 dB, FM = 38.32 dB).

References

[1] Lehpamer, H., *Microwave Transmission Networks–Planning, Design, and Deployment,* second edition, McGraw-Hill, 2010.

[2] Andrew Corporation, Catalog 38, 2000.

[3] Baars, W. M, J., *The Paraboloidal Reflector Antenna in Radio Astronomy and Communication—Theory and Practice,* New York: Springer Science+Business Media, 2007.

[4] Huang, Y., and K. Boyle, *Antennas—From Theory to Practice,* John Wiley & Sons, 2008.

[5] The pamphlet, *The Latest Word in Communications, Microwave New York—Boston System,* Long Lines Department, Plant Division One, 100 William St., New York, 1947.

[6] Mann, M., "The Microwaves Are Coming!" *Popular Science,* November 1947.

[7] Sobol, H., and K. Tomiyasu, "Milestones of Microwaves," *IEEE Transactions on Microwave Theory and Techniques,* Vol. 50, No. 3, March 2002.

[8] Department of Commerce, "Federal Long-Range Spectrum Plan," prepared by Working Group 7 of the Spectrum Planning Subcommittee, September 2000.

[9] Mojoli, L. F., and U. Mengali, "Propagation In Line of Sight Radio Links" (Part II—Multipath Fading), *Telletra Review,* 1988.

[10] Mojoli, L. F., and U. Mengali, "Propagation In Line of Sight Radio Links" (Part I—Visibility, Reflections, Blackout), *Telletra Review,* 1988.

[11] Anderson, D. K., et al., "The Rough Evaporation Duct (RED) Experiment: Microwave and Electro-Optical Transmission Experiments in the Air–Sea Boundary Layer," SPAWAR Systems Center, San Diego, 2002.

[12] ITU, Handbook of Radiometeorology, Geneva, 1996.

[13] Rec. ITU-R P.453-9, "The radio refractive index: its formula and refractivity data," 2003.

[14] Lehpamer, H., *Wind Farms and Microwave Links,* presented at ENTELEC, Houston, Texas, 2013.

[15] Ghobrial, S., and S. Sharif, "Microwave Attenuation and Cross Polarization in Dust Storms," *IEEE Transactions on Antennas and Propagation,* Vol. AP-35, No. 4, 1987.

[16] Williams, E., et al., "The Electrification of Dust-Lofting Gust Fronts (Haboobs) in the Sahel," *Atmospheric Research,* 2008.

[17] Schultz, D. M., "A Review of Cold Fronts with Prefrontal Troughs and Wind Shifts," American Meteorological Society, 2005.

[18] Comparetto, G., "The Impact of Dust and Foliage on Signal Attenuation in the Millimeter Wave Regime," *J. of Space Comm.,* Vol. 11, No. 1, July 1993, pp. 13–20.

[19] Fang, X., et al., "Smart Grid—The New and Improved Power Grid: A Survey," *IEEE Communications Surveys & Tutorials,* Vol. 14, No. 4, Fourth Quarter 2012.

[20] Department of Energy, *Communications Requirements of Smart Grid Technologies,* October 5, 2010.

[21] http://www.google.org/spectrum/whitespace/channel/ (accessed June 05, 2015).

[22] FCC, *Second Memorandum Opinion and Order,* FCC 10-174, Released September 23, 2010.

[23] FCC, *Third Memorandum Opinion and Order,* FCC 12-36, Released April 05, 2012.

[24] Parikh, P., *Investigation of Wireless LAN for IEC 61850 based Smart Distribution Substations,* Doctoral Thesis, School of Graduate and Postdoctoral Studies, The University of Western Ontario London, Ontario, Canada, 2012.

[25] Mason Communications Ltd., "Spectrum Management Strategies for Licence-Exempt Spectrum: Final Report," London, England, November 2001.

[26] Agilent Technologies, *Investigating Bluetooth™ Modules: The First Step in Enabling Your Device with a Wireless Link,* Application Note 1333-2, 2001.

[27] http://www.bluetooth.com/Pages/about-bluetooth-sig.aspx (accessed October 21, 2013).

[28] http://www.bluetooth.com/Pages/Low-Energy.aspx (accessed October 22, 2013).

[29] Shan, Q. et al., "ZigBee Performance in 400 KV Air Insulated Power Substation," *Technological Developments in Education and Automation,* 2010, pp. 15–18.

[30] Ahson, S., and M. Ilyas (eds.), *WiMAX—Technologies, Performance Analysis, and QoS,* Boca Raton, FL: CRC Press, Taylor & Francis Group, 2008.

[31] www.3gpp.org (accessed October 07, 2014).

[32] Tedjini, S., et al., "Antennas for RFID Tags," *Joint sOc-EUSAI Conference*, Grenoble, France, 2005.

[33] Lehpamer, H., *RFID Design Principles,* Second Edition, Norwood, MA: Artech House, 2012.

[34] Lehpamer, H., "Six Steps Towards Privately Owned Radio Networks," *ENTELEC,* Houston, Texas, May 20–22, 2014.

[35] Volakis, J. L., *Antenna Engineering Handbook,* Fourth Edition, NewYork: McGraw-Hill, 2007.

CHAPTER 5

Utility Communications Network Performance and Reliability Considerations

In today's communications networks, with converged voice and data, performance degradation may be as dangerous and costly as hardware failures. Degraded communications channels can result in unacceptable signal transmission quality, loss of information, and dropped connections. High availability does not just mean preventing catastrophic failures; it also means preventing quality and performance degradation.

Communication network *reliability*[1] is the ability of a network to maintain an acceptable level of uptime with minimal mean time between failures. This is often described in terms of *availability*[2] and *survivability.* High availability (and quality) of the communication network is an end-to-end network goal. A *network management system* (NMS) can help identify critical resources, traffic patterns, and performance levels and trending.

Communication network survivability is usually measured in terms of its long-term availability or average network uptime and most organizations expect their network to be continuously available (or at least with as little downtime as possible).

There is an important distinction between terms like *performance, availability,* and *quality.* Although the network is available (meaning operational), network performance can still be poor with increased levels of BER for a number of reasons. In voice networks that may mean reduced sound quality; in data networks, it can cause problems such as constant data retransmissions and incorrect information. In other words, the quality of such a network is low and unsatisfactory.

In addition, increasingly important is *network security,* which means protection of networks and their services from unauthorized modification, destruction, or disclosure. It provides assurance that the network performs its critical functions correctly and that there are no harmful side effects.

---

1. Reliability is "the probability that an item will perform a required function, under stated conditions, for a stated period of time."

2. Availability can be defined as "the proportion of time for which the equipment is able to perform its function." The term "availability" can have a different meaning to different people and in different situations, so it needs to be well defined from start.

5.1 Performance and Communications Channel Protection

5.1.1 Communications System Parameters

A system that is hundreds of miles long will contain, aside from the physical (propagation) media itself, many pieces of equipment along the way. Figure 5.1 depicts a high-level block diagram of the typical communications system [1].

Depending on the application, some of the important parameters that need to be considered when building digital communications systems are availability, latency, resynchronization time, and BER.

Availability is a quotient of the time when the system is considered operational and the total time of observation. Availability of the entire communications circuit depends on many factors, from the physical transmission medium and electronic equipment to human error. It is expressed as a yearly percentage of time [2]. Availability of the system can be increased by using protected systems.

The protective relaying system must be capable of dealing with the *latency* (delays) and *delay time variations* of many digital systems. A difference may exist between delay times in sending a signal versus receiving a signal.

Delay times can also change when alternate signal paths (routing) are used. Current differential relays impose the most stringent delay conditions on communication networks—they typically require delay variations to be less than 0.25 milliseconds.

The system must be designed to be capable of dealing with *interruptions to the communication channel* and a *resynchronization* following a switching operation on the communications network. Simple switching techniques require only a few milliseconds for resynchronization, while others may take tens of milliseconds.

BER is a quotient of the bits received in error and a total number of bits transmitted. Excessive BER, which can cause unwanted tripping of the relays, may be caused by, among other things, high attenuation due to long distances, hardware problems, and interference [3].

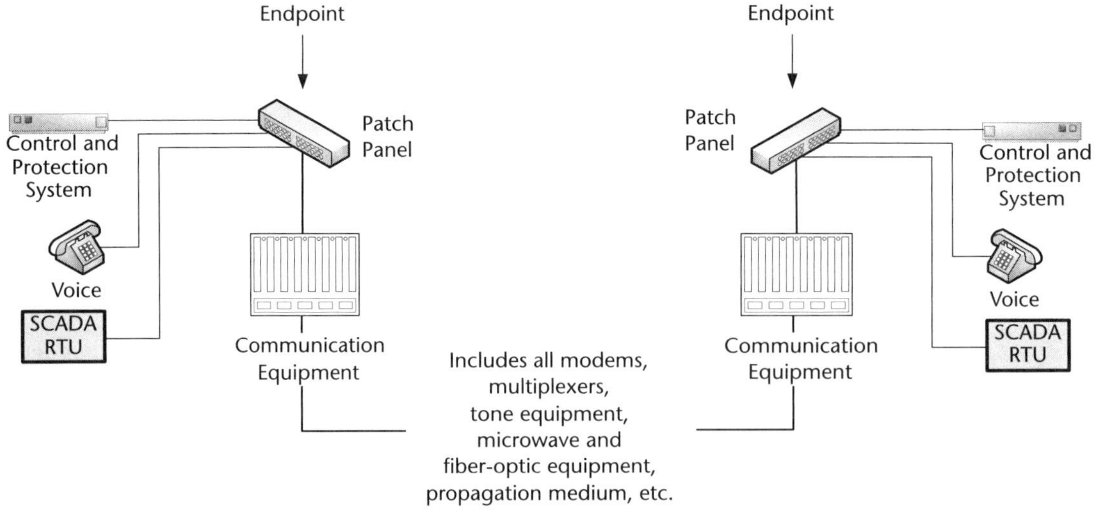

Figure 5.1 End-to-end communications channel.

A few different scenarios for the physical layer of the communications channel are possible: an end-to-end dedicated (owned or leased) optical fiber, a multiplexed network using 64-kbps interfaces (T1 multiplexing, SONET, digital microwave and radio links), and *channel service unit* (CSU) providing 56- or 64-Kbps service over copper lines. Digital microwave can be point-to-point or used as a part of the SONET fiber-optic ring.

An option to use PLC is available but not considered sufficiently reliable to be used for other than SCADA (i.e., noncritical data). PLC, however, is still a popular media for conventional teleprotection such as *direct transfer trip* and *pilot distance protection* [4], although not as a primary but as a secondary communications channel.

With respect to protection relaying communications channels, a dedicated fiber pair has the best digital channel performance. The dedicated fiber pair is virtually immune to electrical interference, has a very low BER, and a very short end-to-end data delay time. Such an application, however, is very expensive and can be prone to long outages if the fiber is cut.

5.1.2 Availability and Redundancy

To improve the availability of the system, one of the methods is to make all the components (or at least the critical ones) of the system redundant. Communications channels that are used for critical traffic must perform during all kinds of power system operations and weather conditions. Redundancy, not only for the equipment but also for the communications path, is often a requirement.

In a redundant configuration, communications systems must remain end-to-end–independent from each other (see Figure 5.2). Typically, there is a primary communications system for protection (and data transmission), as well as a secondary (redundant or backup) system providing path diversity in case the primary

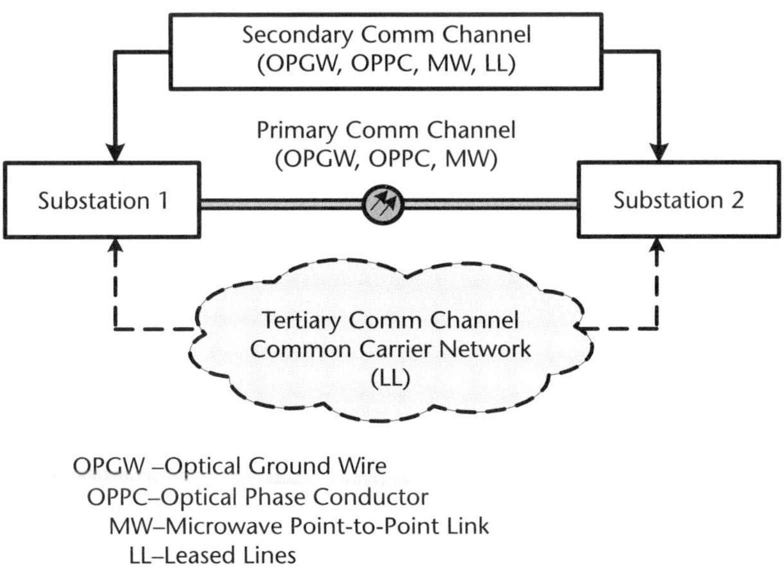

Figure 5.2 Redundant communications system.

system fails [5]. Two geographically separated routes are required to minimize the risk of a simultaneous interruption of both circuits.

The restoration traffic on each route is the duplicate of the working traffic on the other route. Typically, although not always, a dedicated fiber-optic OPGW system is utilized as its primary channel, with a secondary, redundant communications channel provided by another power utility using its own fiber-optic network, or possibly a licensed microwave point-to-point system.[3]

While any owned and operated system's behavior is deterministic in its nature and usually simple to model, leased facilities and services (leased lines) are much more difficult to define since their performance is not under the control of the power utility. So, if there is only one communications system, it has to be designed to the highest performance criteria possible (staying at the same time economically feasible), while three communications systems protecting each other may each might have slightly relaxed performance requirements.

Communication channels will be physically separated to the extent that doing so would meaningfully decrease the probability that a single event or condition would simultaneously disable both communications circuits. In other words, when applicable, the primary and secondary communications systems must be logically and physically segregated into separate racks/panels, with separate entrances to the utility sites. Main and redundant communications circuits should be brought into the power stations with at least 10 feet of separation between the physical paths, ensuring path diversity between the communications systems.

A control center/operations room is also configured with full redundancy, including all the communications equipment. Load sharing and automatic switching of communications servers and communications lines is often required.

Regardless of the transmission network medium and topology, *hardware redundancy* is an option when designing the telecommunications network. Hardware protection types usually employed are 1+1, where one card or module serves as a protection for another one, or N+1, where one card or module protects N other units.

5.1.3 Ring Topology

5.1.3.1 SONET Ring Protection

SONET rings, known as "self-healing rings," use fully duplicated transmission systems and capacity for alternate routing of TDM or STM circuit facilities. The restoration capacity and the associated transmission systems are essentially unused except in the rare occasions of network failure.[4]

Some of the possible types of ring architecture (see Figure 5.3) are two fibers (unidirectional), two fibers (bidirectional), and four fibers (bidirectional).

In *unidirectional path-switched rings* (UPSRs), two redundant (path-level) copies of protected traffic are sent in either direction around a ring. A selector at the

3. If the media on these routes are different (for example, one is fiber optic and the other one is microwave), it is called media diversity. These highly reliable solutions do not come cheap and, in many cases, a compromise between the cost of the network and its deployment time and reliability has to be made. A tertiary (triplicated) communication path is represented with dashed lines because it is not a very common solution.

4. SONET ring can be closed using mixed media (i.e., fiber-optic and microwave).

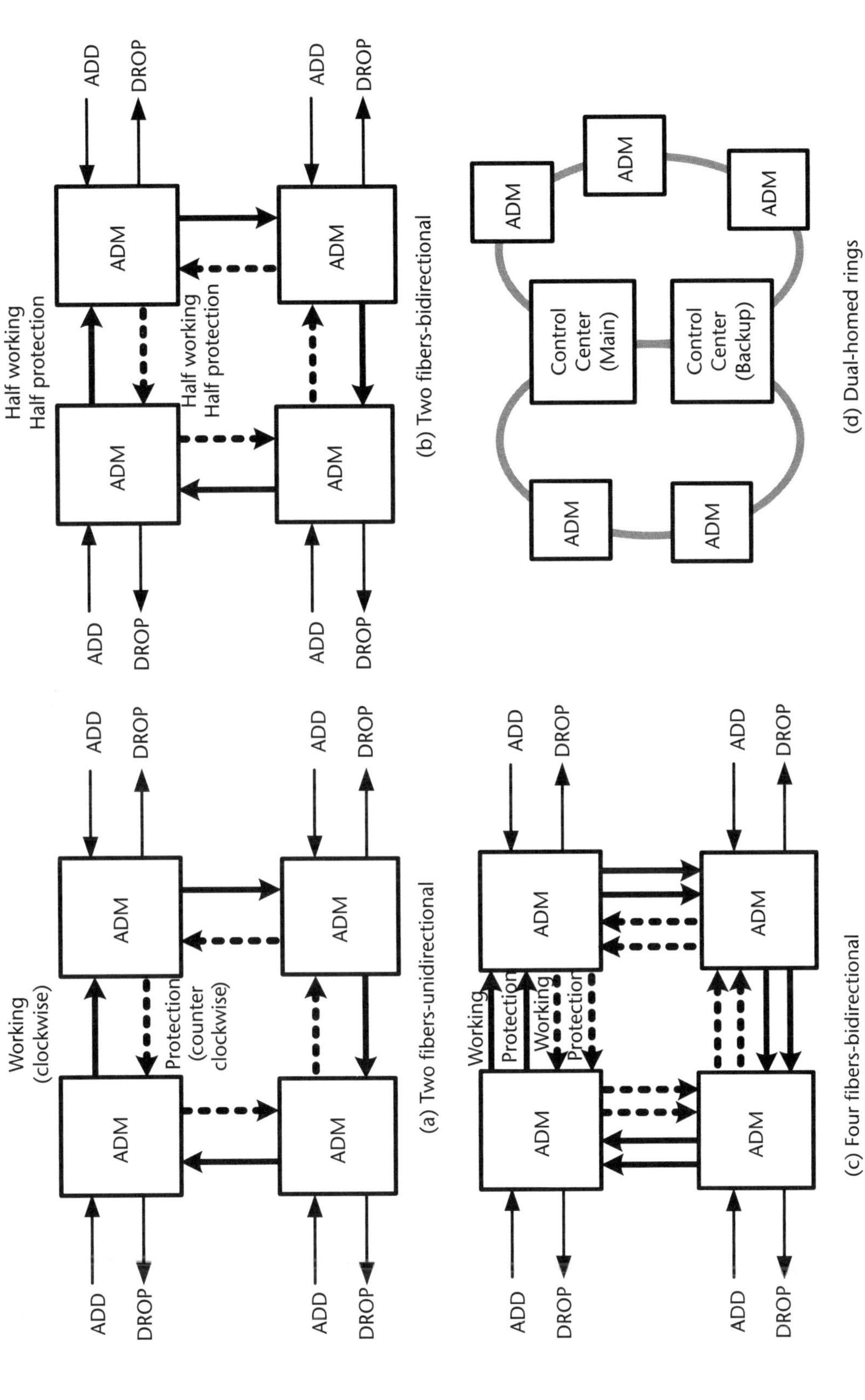

Figure 5.3 SONET ring topology.

egress node determines which copy has the highest quality and uses that copy, thus continuing to function even if one copy deteriorates due to a broken fiber or other failure.

Bidirectional line-switched ring (BLSR) comes in two varieties: two-fiber BLSR and four-fiber BLSR. BLSRs switch at the line layer, and unlike UPSR they do not send redundant copies from ingress to egress. Rather, the ring nodes adjacent to the failure reroute the traffic "the long way" around the ring on the protection fibers. BLSRs trade cost and complexity for bandwidth efficiency, as well as the ability to support additional traffic that can be preempted when a protection switching event occurs. In four-fiber ring, either single-node failures, or multiple-line failures can be supported, since a failure or maintenance action on one line causes the protection fiber connecting two nodes to be used rather than looping it around the ring.

Two adjacent rings might be interconnected via shared node or shared link architecture (*dual-homed ring*),ensuring that traffic is available to pass between adjacent rings at two separate nodes (i.e., since two rings share two nodes, we say that rings have a shared link). If an entire node is lost, the receiving ring equipment will select traffic from the other node. Although it looks expensive, network survivability actually has great potential for cost reduction in the long run. It is typically used for providing redundancy to network operations/control centers or other vital sites.

5.1.3.2 Ethernet Ring Protection

Ethernet Protection Switching Ring (EPSR) enables rings to recover rapidly from link or node failures—within as little as 50 ms, depending on port type and configuration. This is much faster than STP at 30 seconds or even RSTP at 1 to 3 seconds. Ethernet ring protection supports up to 255 ring nodes. The new Recommendation ITU-T G.8032/Y.1344 (2015) defines the automatic protection switching (APS) protocol and protection switching mechanisms for Ethernet ring topologies. Ring protection switching occurs based on the detection of events on the transport entity of each ring link. The events are defined within equipment Recommendations ITU-T G.8021.

The basic idea of an Ethernet ring is to use one specific link to protect the whole ring. This special link is called a *ring protection link* (RPL). If no failure happens in other links of the ring, the RPL blocks the traffic and is not used. The RPL is controlled by a special node called an *RPL owner*, and there is only one RPL owner in a ring; under ring failure conditions, the RPL owner is responsible for unblocking traffic over the RPL.

A ring failure results in protection switching of the RPL traffic. An APS protocol is used to coordinate the protection actions over the ring. Protection switching blocks traffic on the failed link and unblocks the traffic on the RPL. When the failure clears, revertive protection switching blocks traffic over the RPL and unblocks traffic on the link on which the failure is cleared.

5.2 Communications System Latency

Total end-to-end system latency (delay) of the communication system is the sum of the communication media propagation delays and the communication equipment latency. Equipment latency may vary and will be different for different equipment configurations and settings.

5.2.1 Communications Media Propagation Delay

Propagation delay is the length of time it takes for a signal to travel to its destination. For example, in the case of an electric signal, it is the time taken for the signal to travel through a wire. Utilities should carefully consider the selection of communications channels for use with protection relaying systems that perform time-sensitive information, especially when routed through third-party communications providers.

The risks to a given protection system's dependability and security caused by variations in communications channel time delays should also be assessed. If the actual communications channel delay is significantly different in either direction, the average value estimated by the relays may be in error.

A change in delays following a path switch is handled by most modern relays as they automatically measure and adjust for actual channel delay.[5] A pilot wire relay or other relays that do not have this ability will suffer degraded performance and might even cause unintended operation as a result of the current from the remote relay being time-shifted with respect to the local current. *SCADA and telecontrol* applications need dependable and secure communications systems but may tolerate slight delays, depending on the application.

Propagation delay is equal to d/s where d is the distance and s is the wave propagation speed. In radio communications, $s = c$ (i.e. the speed of light). In copper wire, the speed s generally ranges from $0.59c$ to $0.77c$. Table 5.1 summarizes approximate propagation times for different media.[6]

Propagation delay of the entire path can be calculated as:

$$\text{Propagation Delay }[\mu s] = \text{L}[\text{miles}] \times \text{Delay/Mile } \left[\frac{\mu s}{\text{mile}}\right] \quad (5.1)$$

Table 5.1 Media Propagation Times

Distance		*Propagation Time [msec]*		
Miles	*Kilometers*	*Optical Fiber*	*Radio*	*Copper*
0.62	1.00	0.005	0.003	0.004
1	1.61	0.008	0.005	0.007
10	16.09	0.081	0.054	0.070
100	160.93	0.805	0.536	0.700

5. As timing is a key issue, protection relays that have GPS synchronization and delay compensation features will be able to operate correctly even with large delays or delay variations.

6. It is common to use the metric system when it comes to fiber-optic systems, even in the United States and other countries that typically use British units of measurement.

Both the geometry and materials of an insulated conductor affect propagation delay. For typical copper cables, the delay is close to 5 ns/m or 1.5 ns/ft. In most power-system applications this error is not very important, but in systems needing the highest level of performance, it must be corrected. Even very small delays can affect certain applications; for example, for a 250-foot (75-m) cable run, the timing outputs of a GPS clock will be off by up to 375 ns of additional error if there is no compensation for it.

Delays are also caused by routing the output signals over significant distances (e.g., between equipment racks in different buildings).

5.2.2 Equipment Latency

Data about the equipment latency can be obtained from the manufacturer; numbers provided here are just an example. Consider the following calculation of a one-way total end-to-end latency for a 250-mile microwave system:

A microwave system closely following the transmission line route will have a length of approximately 250 miles. The radio propagation medium delay estimation is:

$$\text{Radio Propagation Delay} = 250 \text{ mi } \times 0.0054 \text{ ms/mile} = 1.35 \text{ ms}$$

Assuming an eight-link microwave system with seven repeater sites, each with a 0.3-ms delay, and two terminal sites, with an assumed total 3-ms delay, it is possible to conservatively calculate a worst-case scenario of total end-to-end equipment delay:

$$\text{Microwave Link Equipment Delay} = 3 + 0.3 \times 7 = 5.1 \text{ ms}$$

The total estimated end-to-end microwave link delay, for a 250-mile-long system, is 6.45 ms:

$$\text{Total Microwave Link End-to-End Delay} = 1.35 + 5.1 = 6.45 \text{ ms}$$

5.3 Reliability and Availability Calculations

5.3.1 Mean Time Between Failures (MTBF)

Reliability is defined as the probability that a device will perform its required function under stated conditions for a specific period of time. Estimating the reliability of electronic equipment is generally divided in to two categories: theoretical predictions based on individual failure rates and field-verified reliability based on operation of equipment over time.

Prediction methods are based on component data from a variety of sources: failure analysis, life test data, and device physics. For some calculations (e.g., military applications) *MIL-HDBK-217F* is used, which is considered to be the standard reliability prediction method.

Telecom equipment manufacturers also perform calculations using *Telcordia Technical Reference TR-332*[7], *Reliability Prediction Procedure for Electronic Equipment*. Both of these prediction methods have several assumptions in common (e.g. constant failure rate, the use of thermal and stress acceleration factors, quality factors, and use conditions).

Reliability is quantified as MTBF for repairable products and *mean time to failure* (MTTF) for nonrepairable products.

A correct understanding of MTBF is important. For example, it does not mean that a device with an MTBF of 100,000 hours will last for an average of 100,000 hours; according to the theory behind the statistics of confidence intervals, after a time equal to the MTBF only 37% of the units will still be working.[8]

For a product with an MTBF of 100,000 hours, and an operating time of interest of, for example, five years (43,800 hours), we can say that:

$$P(t) = e^{-t/MTBF} = e^{-43.800/100,000} = 0.645 \tag{5.2}$$

This means that there is a 64.5 percent probability that this particular product will operate five years without a failure, or that 64.5 percent of the units in the field will still be working after five years.

The *failures in time* (FIT) is a number of failures in 10^9 hours (114,155 years), which is equivalent to MTBF of 10^9 hours.

$$FIT = \frac{10^9}{MTBF} = 10^9 \cdot \lambda \tag{5.3}$$

For electronic products, it is commonly assumed that during the useful operating life period the parts have constant failure rates and that part failure rates follow an exponential law of distribution. If the MTBF is known, one can calculate the *failure rate* (λ), expressed as number of failures per hour, as the inverse of the MTBF. The formula for λ is:

$$\lambda = \frac{1}{MTBF} \tag{5.4}$$

5.3.2 Availability Calculations

Availability (A) can be defined as the proportion of time for which the equipment is able to perform its function:

7. The Telcordia (originally known as Bellcore) method is based on Mil-HDBK 217 but is tailored for the telecom industry. The last revision was issued in January of 2011 and is known as SR-332 issue 3. The following documents were fully or partly replaced by SR-332: TR-332, TR-NWT-000332, TA-000-23620-83-01, and IP-10425. Other related Telcordia documents of interest are GR-357, GR-2969, SR-1171, and SR-TSY-001369.

8. MTBF is a measure of reliability, but it does not mean the expected life, the useful life, or the average life. MTBF is a statistical measure, and as such, it cannot predict anything for a single unit.

$$A = \frac{MTBF}{MTTR + MTBF} \tag{5.5}$$

Availability is different from reliability in that it takes repair time into account. An item of equipment may not be very reliable, but if it can be repaired quickly when it fails, its availability could be high. Maintenance can be corrective (repair) or preventative (to reduce the likelihood of failure).

Mean time to repair (MTTR)[9] is defined as the total amount of time spent performing all corrective or preventative maintenance repairs divided by the total number of those repairs. It is the expected span of time from a failure (or shut down) to the repair or maintenance completion.

Unavailability (U) can be calculated as:

$$U = 1 - A \tag{5.6}$$

For n pieces of equipment connected in series (tandem):

$$U = U_1 + U_2 + \ldots + U_n \tag{5.7}$$

For n pieces of equipment connected in parallel:

$$U = U_1 \cdot U_2 \cdot \ldots \cdot U_n \tag{5.8}$$

Mean time to restore (MTR)[10] is calculated as follows:

$$\text{MTR} = \text{MTTR} + \text{Travel Time} + \text{Availability of Spares} = \frac{1}{\mu} \tag{5.9}$$

Availability of Spares = $(1 - P)T_S$, where P is the probability that a spare module is available when needed, and T_S is time to obtain the spare module.

$$MTR = \frac{1}{\mu} \tag{5.10}$$

where

μ: Restore rate per hour.

Probability of system working (P_0):

---

9. Sometimes Mean Down Time (MDT) is used in this formula instead of MTTR. The reason is that MTTR may not be the same as MDT because the failure may not be noticed for some time after it has occurred or it may be decided not to repair the equipment immediately or the equipment may not be put back in service immediately it is repaired. Whether MDT or MTTR is used, it is important that it reflects the total time for which the equipment is unavailable for service; otherwise, the calculated availability will be incorrect.

10. Sometimes MTTR and MTR are used interchangeably, so it is important to understand what given value(s) are comprised of.

$$P_0 = \frac{\mu}{\mu + \lambda}$$

Probability of system failed (P_1):

$$P_1 = \frac{\lambda}{\mu + \lambda} \tag{5.12}$$

and

$$P_0 + P_1 = 1 \tag{5.13}$$

One year has 8,766 hours (i.e., 525,960 minutes), so downtime (*DT*) in [min/year] can be calculated as:

$$DT = P_1 \cdot 525{,}950 \tag{5.14}$$

5.3.3 Availability of Fiber-Optic Systems

The model and methodology described here uses failures in time (FIT), from which failure rate (λ) can be calculated, and MTR, from which repair rate (μ) can be calculated [6]. A failure rate of one failure in 10^9 hours equals 1 FIT.

The model presented in Figure 5.4 utilizes two different values for MTR; fiber-optic restoration MTRs are, typically, longer than communications equipment MTRs that are based on the replacement of faulty cards. An MTR of eight hours (or less) is typical for communications equipment inside a control room. Individual utilities often define MTR based on the availability of spare equipment and access to the site under a worst-case scenario.

Typical fiber-optic MTR is in the range of 12–24 hours. Although not shown in the Figure 5.4, from a communications channel availability perspective, site power

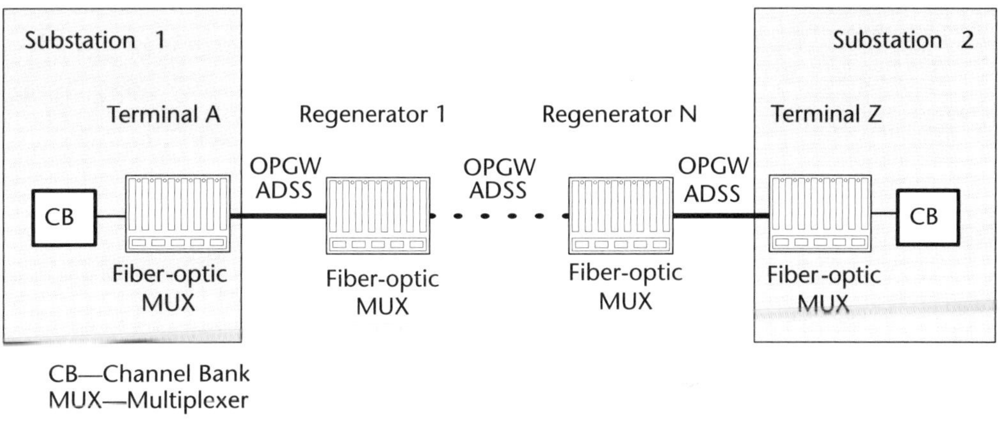

Figure 5.4 Fiber-optic communications channel block diagram.

(rectifiers and the generator) is connected in series with the rest of the equipment and fiber-optic cable.

Fiber-optic MTR is an important parameter because circuit availability calculations show a high level of dependence on fiber-optic MTR; the values should be based upon the individual utility's fiber-optic restoration experiences and restoration programs in place. MTR includes incidents where service was restored by switching service to working (spare) dark fiber(s), as well as complete fiber-optic restorations due to cable cuts.

The use of temporary cables and temporary mechanical splices can reduce restoration time in the case of complete cable failures. In addition, not all equipment failures will affect traffic.

5.3.3.1 Fiber-Optic Cable Downtime

The recommended fiber-optic failure rate is 342 FIT/mile (212.50 FIT/km), which equates to three fiber-optic cable failures per 1,000 route miles per year, or 1.86 fiber-optic cable failures per 1,000 route km per year. This failure rate is based upon telecom industry studies of fiber-optic sheath failure rates. The recommendation is conservative because not all fiber-optic sheath failures result in service affecting outages (i.e., damage to lit fibers).

Data from Telcordia shows that aerial fiber cable is more reliable than underground cable and that OPGW is the most reliable aerial cable. *Network Reliability Steering Committee* (NRSC)[11] data for all network categories of "facility failure" indicate that underground cable damage accounts for approximately 72 percent of incidents, compared to 6 percent for aerial cable.

Service-affecting failure rate data for both ADSS and OPGW cables reveal them to be more than 18 times more reliable than conventional in-ground systems. In many telecommunications systems, the goal for maximum service-affecting failures is 2.5 failures per 1,000 kilometers per year.

Based on the 2001 analysis [7], aerial fiber-optic cables are expected to provide the longest continuous telecommunications systems service of all cables studied. OPGW and ADSS cables have more than 25 times the in-service reliability of buried fiber-optic cables. The calculations show that OPGW has 0.14 failures per 1,000 route miles per year (or approximately 16 FIT/mile.)

5.3.3.2 Fiber-Optic Equipment Downtime

When it comes to individual pieces of electronic equipment, Telcordia requirements state that the downtime of a two-way channel within a SONET multiplexer, due to hardware failure, must be less than 1.75 min/year in a CO environment and 5.25 min/year in an outside plant (OSP) environment [8]. The corresponding objectives for these parameters are 0.25 min/year in a CO environment and 0.75 min/year in an OSP environment.

11. A Network Reliability Steering Committee (NSRC) was formed under the auspices of the Alliance for Telecommunications Industry Solutions (ATIS) to monitor network reliability utilizing major outage reports. NRSC is composed of representatives from major wireline and wireless carriers and suppliers of telecommunications equipments and they meet quarterly to discuss trends in network reliability.

These requirements and objectives apply to all system elements needed to process a two-way channel, including the core system as well as the high-speed and low-speed interfaces. System-reliability analysis employing Markov modeling is used to determine system downtimes. As specified in Telcordia, GR-418-CORE, this analysis assumes an MTR of two hours for the CO environment and four hours for the OSP environment.

Individual *field replaceable unit* (FRU) failure rates should be determined using the method described in Telcordia documents [9]. FRUs are system elements that can be replaced in the field, including items such as circuit packs and removable optical interfaces.

Software and procedural downtime should be included in the availability calculations, so 30 minutes is usually assumed. Forecasting the contribution of software and hardware upgrades and replacements, as well as procedural errors to system downtime is subjective, and quite often hard to predict.

5.3.3.3 Fiber-Optic System Availability Calculation

In this example we will calculate end-to-end availability of a 200-mile-long fiber-optic route, mostly OPGW, and three optical regeneration stations (i.e., every 50 miles). For simplicity, we will assume that all fiber-optic multiplexers have the same FIT (which may not be the case in real projects) and that spares are available 100% of the time when needed.

We will use MTR of eight hours for all electronics and 12 hours for the fiber-optic cables. Battery backup and diesel generators are included in the reliability of site electronic equipment.

- Terminal site = 10,000 FIT (42.07 min/year);
- Regenerator site = 2,500 FIT (10.52 min/year);
- Equipment total = 27,500 FIT (115.73 min/year);
- Fiber-optic cable = 3,200 FIT (20.20 min/year);
- Total downtime = equipment + FO cable + maintenance = 115.73 + 20.20 + 30 = 165.93 min/year;
- End-to-end availability = 99.9685 %.

In this calculation, only 30 minutes was allotted for procedural errors and unscheduled outages due to human error, and experience teaches us that merely one serious operator's error could jeopardize the calculated and predicted outage times. In other words, a high reliability of equipment is only one link in the high availability chain, consisting of equipment, OPGW cable and a well-trained operations and maintenance staff.

This calculation is a simplified example of point-to-point linear communications channel availability.[12] Using redundant equipment and a redundant physical

12. Sometimes an end-to-end combination of privately owned fiber-optic and/or microwave systems and leased lines connected in series (i.e., hybrid system) is necessary. In that case, leased lines and their unavailability is the major contributor to the overall end-to-end system availability.

path (or ring configuration) will significantly increase overall availability of the telecommunication system.

5.4 Communication Requirements for Protection Relaying

5.4.1 Redundancy

The reliability of the bulk electric system (BES) is normally measured by determining the performance of all the various power system elements and their ancillary systems. Protection systems, being important ancillary systems, must support the overall reliability requirements of the BES. The North American Electric Reliability Corporation (NERC) reliability standards define the level of reliability to which each owner must design the BES, and this in turn, can be used to determine the performance requirements of electric system elements such as breakers, relays, and protection systems [10].

Protection relays monitor critical electrical and mechanical quantities and initiate emergency shutdown when they detect out-of-limit conditions. Protection relays must detect abnormal conditions, shut down appropriate equipment, and must not incorrectly operate and unnecessarily shut down equipment at any other time.

Planning a teleprotection communications channel requires certain considerations, such as availability of a communications path and selecting the medium, choosing the appropriate communications equipment to connect to the path, compatibility between those devices and the interfacing relays, timing and protocol issues, system security (should not operate unnecessarily), and dependability (must operate when required) [11].

A view of the typical protection relay chassis, showing redundant power and communication modules is shown in Figure 5.5.

Redundancy of the components is one way to increase the reliability of the system. Redundancy is a system design approach that duplicates components and/or systems to provide alternatives in case one component and/or system fails. To ensure the protection system redundancy, the following components should be doubled: AC current source, AC voltage source, protective relay, communication channel, DC circuitry, auxiliary trip relay, breaker trip coil, and station DC source (Figure 5.6).

Fully independent communications channels are the hardest, and sometimes the most expensive, elements to provide for protection system redundancy. Redundant communications equipment provides functionally equivalent protection, so no immediate action is required, although repair or replacement will have to be made as soon as possible.

The purpose of the communications channel in a distance protection schemes is to transmit information about the system conditions from one end of the protected line to the other, including requests to initiate trip (transfer tripping scheme) or prevent tripping (blocking scheme) of the remote circuit breaker. The terminology of the various schemes varies widely, according to local custom and practice.

The most desirable protection scheme is a combination of two arrangements (i.e., instantaneous tripping over the whole line length plus back-up protection to adjacent lines). This can be achieved by interconnecting the distance protection

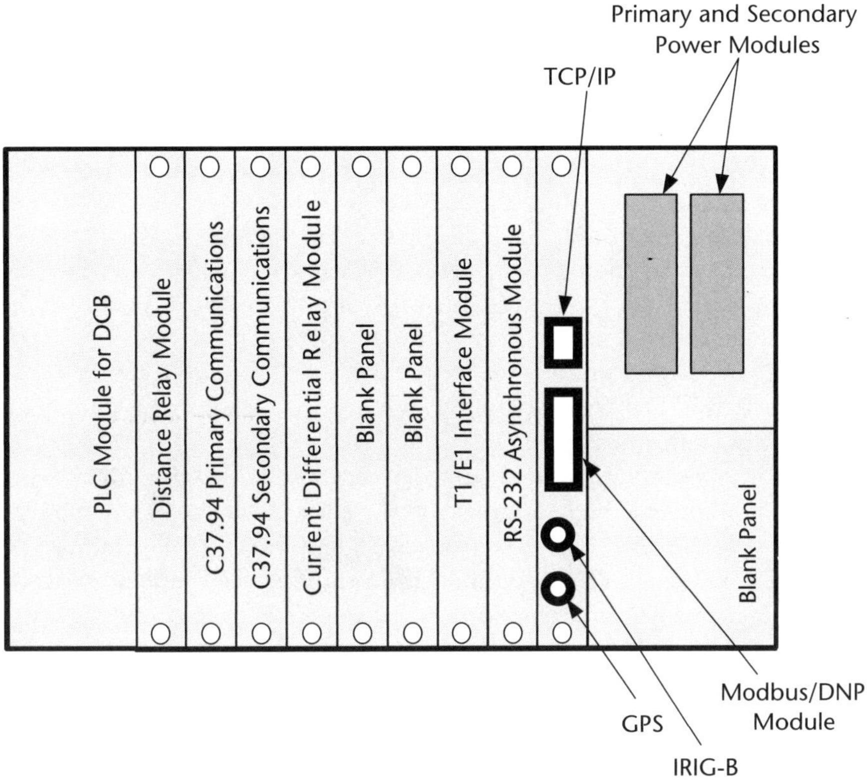

Figure 5.5 Protection relay modules.

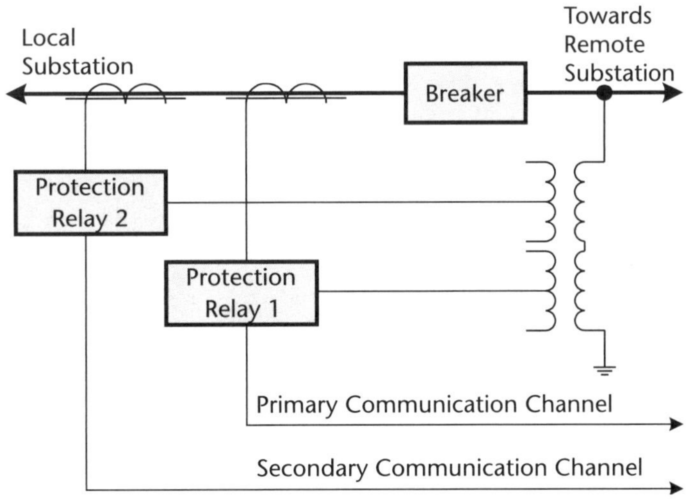

Figure 5.6 Dual control and protection system.

relays at each end of the protected feeder by a communications channel [12]. Other protection schemes are also possible, and each utility has its own way to protect its assets and prevent large-scale electrical outages.

5.4.2 Protection Schemes and Communications Channels

This section provides some basic information on communications channel requirements for different protection schemes. Details of relay protection and their operation are out of the scope of this book but can easily be found in the literature [13].

5.4.2.1 Current Differential Scheme

In a current differential scheme, a true differential measurement is made. Data from the relay at the remote terminal is provided to the local relay using a communications channel. This data is utilized by the local relay to develop a representation of the magnitude and phase angle of the currents, measured by the remote relay. This representation is then compared to the magnitude and phase angle of the currents, measured by the local relay.

From a communications perspective, there are two main types of current differential relaying. The first combines the currents at each terminal into a composite signal and compares these composite signals through a communications channel to determine if a fault is present within the line section. The second type samples individual phase currents, converts each current to a digital signal, and transmits these signals between terminals using a wide-band channel to determine if a fault is present within the line section.

5.4.2.2 AC Pilot-Wire Relays

Traditional pilot-wire relays use a metallic pilot-wire pair for the communications medium. These relays communicate via power system frequency signals on the pilot-wire. Pilot-wire relays are traditionally used on short lines and are easy to apply.

It is possible to use pilot wires for the digital communications channel for protective relaying purpose. The performance of a digital communications channel may be affected by the interference and noise from power cables and other objects.

AC pilot wire relays were designed with no consideration for channel delay because they use a metallic pilot wire as communications link. Special care needs to be taken when applying these relays over a digital communications network. The relays are not able to tolerate any asymmetrical delay. End-to-end delay must be kept very short in order to maintain reliable operation, typically less than 1 ms.

5.4.2.3 Phase Comparison Schemes

A phase comparison scheme compares the phase angle of the fault current at the two terminals of the protected line. If the two currents are essentially equal and 180 degrees out of phase, the relays detect an external fault and do not initiate a trip. If these two currents are essentially in phase, the relays detect it as an internal fault and initiate a trip to the appropriate breakers.

5.4.2.4 Directional Comparison Schemes

Directional comparison schemes frequently use distance relay functions that are inherently directional and directional ground over-current relays.

For *direct under-reaching transfer trip* (DUTT), *permissive under-reaching transfer trip* (PUTT), and *permissive over-reaching transfer trip* (POTT) schemes, communications channel delays simply delay the trip output; and changes in communications channel delay because of switching or rerouting only affect the tripping delay. Loss of communications disables the scheme, requiring some form of backup protection to affect a trip. Permissive transfer trip schemes typically require trip times less than 12 ms.

For *directional comparison blocking* (DCB) schemes, communications channel delay is critical to performance of the DCB. The coordinating time delay must be set greater than the maximum expected channel communications delay to prevent tripping for external faults. Therefore, switched communications schemes are not well-suited for use with DCB schemes. The DCB scheme is not dependent on the communications channels for tripping, only for blocking trips for faults outside of the zone of protection (i.e., the protected line). If the communications channel is lost, one terminal of the DCB scheme may trip for a fault beyond the remote terminal.

5.5 Cyber and Grid Security

Threats to the electric system range from weather-related incidents to terrorist attacks—including physical attacks as well as attacks on computer systems, or cyberattacks [14]. The main risk from weather-related damage or a terrorist attack against the electric power industry is a widespread power outage that lasts for an extended period of time.[13]

Cybersecurity refers to all the security issues in automation and communications that affect any functions related to the operation of the electric power systems, meaning the following:

- Data cannot be altered undetectably (integrity).
- Communications parties involved must be validated as genuine (authentication).
- Only requests and commands from authorized users can be accepted by the system (authorization).
- Data must not be accessible to any unauthenticated users (confidentiality).

5.5.1 Challenges of Today's Communication Networks

Many legacy protocols are designed around a simple principle of request and respond and lack authentication, encryption, or other basic security measures, so they

13. Of the transmission system's physical infrastructure, high-voltage (HV) transformers are the most critical component. Utilities rarely experience loss of an individual HV transformer, but recovery from such a loss takes months if no spare is available. On the other hand, utilities regularly experience damage to transmission towers due to both weather and malicious activities and are able to recover from this damage fairly rapidly as they generally have not been widespread or long-lasting.

may be vulnerable to network-based attacks. Security concerns common to most industrial control, including utility communication protocols, include the following:

- Protocol commands may be available that are capable of forcing slave devices into inoperable states, including powering off devices, forcing them into a "listen only" state, disabling alarming.
- Network or transport errors (e.g., malformed packets or excessive latency) can cause protocol failure.
- Protocol commands may be available that are capable of restarting communications and otherwise interrupting processes, and clearing, erasing, or resetting diagnostic information such as counters and diagnostic registers.
- Protocol commands may be available that are capable of requesting sensitive information about the controllers, their configurations, or other need-to-know information.
- Most protocols are application layer protocols transported over TCP; therefore it is easy to transport commands over nonstandard ports or inject commands into authorized traffic flows.
- Protocol commands may be available that are capable of broadcasting messages to many devices at once (i.e., a potential denial of service).
- Protocol commands may be available to query the device network to obtain defined points and their values (i.e., a configuration scan).
- Protocol commands may be available that will list all available function codes (i.e., a function scan).

The attention on SG security has been focused mostly on cyberthreats instead of physical threats because the consequences of a cyberattack on the grid telecommunications network can be catastrophic. Simple injection of malicious protocol commands can provide control over the target process. Altering legitimate protocol traffic can also alter information about a process and disrupt the legitimate controls that are in place over that process. A man-in-the-middle attack could provide both control over a process and misrepresentation of data back to operator consoles.

By analyzing the security level of each substation and importance in a power grid, an intruder may find the optimal combination of target substation(s) that can trigger a sequence of *cascading events*, leading to a system blackout. Therefore, the impact of simultaneous cyber (and/or physical) attacks to multiple substations can be much higher than that of a single substation attack. [15].

5.5.2 FERC and NERC

System downtime, data loss, and facility control breakdowns quickly become business critical issues for power utilities and their customers. The Energy Policy Act of 2005 (Energy Policy Act) gave the Federal Energy Regulatory Commission (FERC) authority to oversee the reliability of the bulk power system, commonly referred to as the bulk electric system or the power grid of North America and Canada. This includes the authority to approve mandatory cybersecurity reliability standards [16].

NERC, which FERC has certified as the nation's *electric reliability organization*, developed *critical infrastructure protection* (CIP) cybersecurity reliability standards. On January 18, 2008, the commission issued Order No. 706, the final rule approving the CIP (CIP 002 through 009) reliability standards, while concurrently directing NERC to develop significant modifications addressing specific concerns.

Additionally, the electric industry is incorporating *information technology* (IT) systems into its operations—commonly referred to as smart grid—as part of nationwide efforts to improve reliability and efficiency. There is concern that if these efforts are not implemented securely, the electric grid could become more vulnerable to attacks and loss of service. To address this concern, the *Energy Independence and Security Act* of 2007 (EISA) gave FERC and NIST responsibilities related to coordinating the development and adoption of smart grid guidelines and standards.

Power utilities are subject to regulatory security requirements that are mandated by the NERC. Compliance with NERC-CIP standards requires comprehensive cybersecurity solutions including segmentation, authentication, authorization, monitoring, logging, and training and comprehensive physical security solutions including access control and video surveillance. NERC violations may lead to costly sanctions and remedial action directives that must be immediately addressed.[14] Moreover, NERC can assess fines of up to a million dollars a day per violation.

The challenge for bulk electric systems is to meet (or exceed) security compliance standards to prevent potential attacks on the network, thereby contributing to grid reliability, while balancing performance and *total cost of ownership* (TCO). Power utilities need to be vigilant against both intentional and unintentional insider threats from current and former employees and contractors, as well as outsider threats from hackers and cyberterrorists. In addition to issues for operational data, there may also be database vulnerabilities in customer information, billing, and financial systems.

NERC offers a consistent framework for security control perimeters and access management with incident reporting and recovery for *critical cyber assets* (CCAs). See Table 5.2 for the latest revision of the guidelines.[15]

5.5.3 Security of Utility Networks

As mentioned earlier, network security is a protection of networks and their services from unauthorized modification, destruction, or disclosure. It provides assurance that the network performs its critical functions correctly and that there are no harmful side effects.

14. With oversight by the U.S. Federal Energy Regulatory Commission (FERC) and Canada's National Energy Board (NEB), cybersecurity standard compliance has become an increasing priority for the power utility industry. FERC enforcement of the NERC guidelines also applies to Canadian power generators wishing to export power to the United States.

15. Revision 5 is the latest revision at the time of writing this text. CIP-010-2, CIP-011-2, and CIP-014-1 are the new additions to the standards.

Table 5.2 NERC Critical Infrastructure Protection

CIP	*Description*	*Summary*
NERC CIP-002-5.1	Critical Cyber Asset Identification	Identification and documentation of risk-based assessment methodology used to identify CCAs.
NERC CIP-003-5	Security Management Controls	Documentation and implementation of cybersecurity policy reflecting commitment and ability to secure CCAs.
NERC CIP-004-5	Personnel and Training	Maintenance and documentation of security awareness programs to ensure personnel reinforcement on sound security practices.
NERC CIP-005-5	Electronic Security Perimeter(s)	Identification of electronic security perimeters surrounding CCAs and all access points therein.
NERC CIP-006-5	Physical Security of Critical Cyber Assets	Creation and maintenance of a physical security plan, including processes, tools, and procedures to monitor perimeter access.
NERC CIP-007-5	Systems Security Management	Ensuring that new cyber assets and significant changes to existing cyber assets within the electronic security perimeter do not adversely affect existing cybersecurity controls.
NERC CIP-008-5	Incident Reporting and Response Planning	Develop and maintain a cybersecurity incident response plan that addresses classification, response actions and reporting.
NERC CIP-009-5	Recovery Plans for Critical Cyber Assets	Creation and review of recovery plans for CCAs.
NERC CIP-010-2	Configuration Management and Vulnerability Assessment	To prevent and detect unauthorized changes to CCAs.
NERC CIP-011-2	Information Protection	To prevent unauthorized access to CCAs information by specifying information protection requirements.
NERC CIP-014-1	Physical Security	To identify and protect CCAs and their associated primary control centers, that if rendered inoperable or damaged as a result of a physical attack could result in widespread instability, uncontrolled separation, or cascading within an interconnection.

5.5.3.1 Operations and Corporate Networks

SCADA and/or protection systems (i.e., operations networks) control some of the most critical national infrastructure, including nuclear power stations, electric transmission and distribution facilities, water pumping and waste treatment plants, oil processing facilities, and chemical plants. Loss of access to or misuse of these systems could result in severe physical damage, disruption and financial loss to a company and their customers and partners. Therefore, security of these operations networks should be a high priority.[16]

Traditionally, utility operations networks have been separated from other corporate networks to minimize exposure to unsecure areas, such as the Internet. Recently however, more organizations are connecting operations networks with other potentially unsecure networks in order to cut costs, share operational information,

16. Most SCADA protocols were never intended for use on publically accessible networks, and in some cases, not even on IP networks. MODBUS, a common SCADA protocol, was originally designed for use only within simple process control networks to enable low speed serial communications between clients and servers. In order to communicate on an IP network, MODBUS TCP merely encapsulates serial data within a TCP packet before sending it across the network. No additional protection has been added to secure communications sent using the protocol.

or distribute ordering/billing data. Even when connecting operations networks to other networks is prohibited by corporate policy, incorrectly installed systems can unintentionally bridge networks together—putting operations networks and the processes they control at risk.[17]

Generally speaking, NERC CIP-005 Electronic Security Perimeter(s) is applicable to network security. According to NERC CIP-005, at a minimum, all network connections across the perimeter must have a firewall properly configured so only authorized traffic and connections are permitted within the secured zone. This requirement includes IP address-based packet filtering, TCP and UDP transport layer inspection, and review of application layer traffic such as Telnet, HTTP, and HTTPS.

Accountability and *auditability* are key elements of all NERC requirements. Thus, network security reporting and monitoring at multiple layers and on an end-to-end basis are important components of any compliancy program (e.g., logging and alerting for periodic audits or real-time analysis)—especially when IP protocols are at issue.

Most network and computer security controls have been developed for IT systems, which do not have the same operating and maintenance characteristics as SCADA systems. SCADA systems have to be available all day, everyday. Furthermore, they cannot be shut down for running back-ups, have to have deterministic response times, should not accidentally lock out operators in crisis situations, do not usually have excess computing capacity, and might have life and death consequences if malfunctions occur [17].

Often, security for SCADA systems consists of nothing more than a basic password that is passed in plain text from the control system to the RTUs. These unprotected passwords can easily be intercepted by malware and packet sniffers, and unfortunately these passwords quite often remain set to the manufacturer's default value.

Although corporate IT policies require password changes on a monthly basis, this is rarely applied to SCADA systems. This is due to the resources needed to manually change passwords on potentially hundreds of devices and the risk to the company in case these devices become inaccessible.

The widespread use of encryption is new and not very common to the electric power industry. There are proven and reliable encryption systems designed for industrial use, but the majority of SCADA systems in use do not apply any authentication or encryption methods.

The user interface is another area where SCADA and IT requirements differ. Corporate IT networks carry a large amount of data required for the daily functioning of corporations (i.e., e-mails and web pages that connect employees internally among themselves and externally with the public and partners). Too much information, such as graphics, schematics, and large amounts of raw data on SCA-

17. Following the September 11th terrorist attacks, these concerns prompted eminent U.S. scientists to write in 2002 an open letter to President Bush in which they warn: "The critical infrastructure of the United States, including electric power, finance, telecommunications, health care, transportation, water, defense and the Internet, is highly vulnerable to cyber attack. Fast and resolute mitigating action is needed to avoid national disaster." http://www.pbs.org/wgbh/pages/frontline/shows/cyberwar/etc/letter.html (accessed Dec 26, 2014).

DA screens, tend to overwhelm the operator and make it difficult to detect critical situations in a timely fashion.

Many utilities are still trying to improve security of operations networks by keeping them separate from their corporate IT networks.

5.5.3.2 Routable and Nonroutable Protocols

From a communications perspective, all communications protocols have at their core a "payload" of information that is to be transported. That payload is then wrapped in either a simple addressing and error-detection envelope and sent over a communication channel [i.e., traditional (nonroutable) serial protocols] or it is wrapped in additional layers of application layer and networking protocols that allow transport over wide area networks (routable object-oriented protocols like IEC 61850).

A *nonroutable protocol* is a communications protocol that contains only a device address and not a network address for sending data from one network to another. Nonroutable protocols operate in their own network, and they are designed in such a way to think that all computers with which they communicate are on the same network as the source computer. Examples of nonroutable protocols are NetBIOS and DEC's LAT protocols.

A *routable protocol* is a communications protocol that contains a network address as well as a device address. It allows packets to be forwarded from one network to another. Examples of routable protocols are TCP/IP, IPX, AppleTalk, SNA, XNS, and DECnet.

Although NERC specifies that "Direct serial, non-routable connections are not included (typically leased line communications) as there is no perimeter or firewall type security that should be universally mandated across all entities and all serial communication situations," the nonroutable communication links (dial-up modems and other serial links) deserve consideration as part of any comprehensive security plan. Some potential protections should include the following:

- Ensuring that all devices involved in such communications are protected from unauthorized physical access.
- Considering encryption or other protections for communication paths. Encryption should be avoided, however, if the data latency introduced by the encryption process could cause the intended operation to fail (protective relaying schemes, for example, where millisecond timing is needed to protect the equipment from damage).
- Providing additional monitoring and/or isolation for devices with nonroutable communications links to potentially untrusted devices.

It is not always obvious what constitutes a device connected through the "routable connection." For example, consider the following:

- Question: For protection relays that are serially connected to a router/protocol converter, where the router/protocol converter has external routable connectivity, are the relays themselves considered cyberassets that can be directly accessed through external routable connectivity (ERC)?

- Answer: A protocol converter cannot be used to avoid compliance. It doesn't matter if relays are serially connected to the router. If the relay can be accessed outside of its associated electronic security perimeter and its state can be changed through any means by using a "bidirectional routable protocol connection," then it is considered to have ERC.

5.5.3.3 Physical Security and Surveillance

Security and surveillance is a necessity in utility applications. Physical security addresses design, implementation, and maintenance of countermeasures that protect the physical resources of an organization. Most technology-based controls can be circumvented if an attacker gains physical access to the devices being controlled. As a result, physical security should receive as much attention as cybersecurity in the security development process.[18]

The topic of physical security at electric substations has received increased attention from the media, politicians, and regulators. FERC on March 7, 2014, ordered owners and operators of the bulk-power system to take at least three steps to protect physical security: (1) perform risk assessment of their substations and primary control centers; (2) evaluate potential threats and vulnerabilities to those facilities, develop/document, and implement a security plan to address potential threats and vulnerabilities; and (3) provide third-party verification.[19]

Limiting access to control rooms, monitoring facility grounds, and establishing the ability to record events is required in order to protect assets, ensure safe operation, and guard against malicious acts. Figure 5.7 shows some of the devices commonly used in surveillance systems.

Pan/tilt/zoom (PTZ) dome cameras have become the most popular choice for applications that require the ability to position and zoom the camera to a specific area of interest. Advanced low-light technology produces color images even under extreme low-light conditions, so users can now monitor scenes at night and in dark areas.

True substation security means stopping an event by detecting intruders as early as possible—ideally before they enter a secured area—and alerting security staff. When it comes to securing substations, real-time awareness is critical for preventing copper theft, vandalism, accidents, and other activities that could dis-

18. Grid cyber security is a hot issue inside the utility industry. However, report on an attack on the Metcalf power station in San Jose, CA on April 16, 2013 highlighted physical security vulnerabilities. The attack was a simple brute force physical assault on power grid infrastructure which Jon Wellinghoff, former chairman of the Federal Energy Regulatory Commission described as "the most significant incident of domestic terrorism involving the grid that has ever occurred". The assailants cut phone lines in an underground vault then began shooting at transformers with weapons that used shell casings similar to an AK-47. According to the account, cameras were aimed inside the substation perimeter and therefore did not capture images of the assailants. No arrests have been made. Grid operators avoided a blackout by calling on power plants to ramp up generation and by rerouting power around the site.

19. More information can be found in the latest release of NERC CIP-14-1 Physical Security.

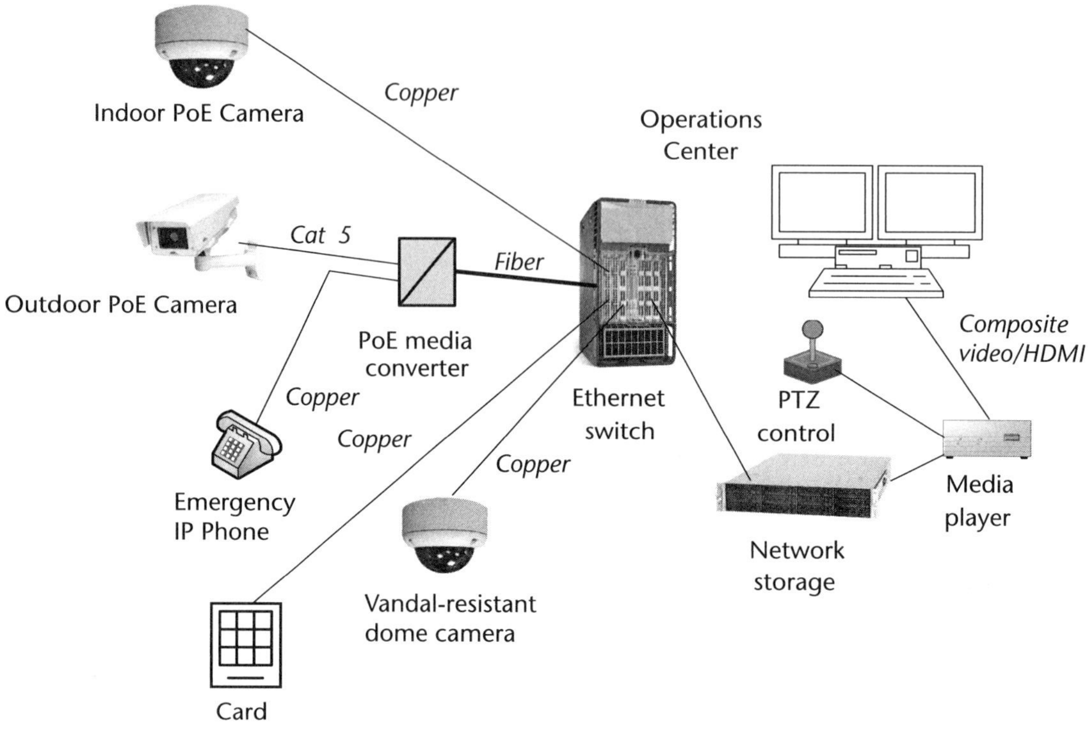

Figure 5.7 Surveillance equipment.

rupt the stability of the power grid. Relying on *video analytics*[20] to automatically monitor surveillance cameras and alert for events of interest is in many cases much more effective than reliance on a human operator. Video analytics applications can perform a variety of tasks ranging from real-time analysis of video for immediate detection of events of interest, to analysis of prerecorded video for the purpose of extracting events and data from the recorded video.

A military-grade technology that became available for mass market recently (mainly due to reduction in price) is *thermal imaging*. Thermal imaging cameras see heat and not light and therefore cannot be blinded by natural or other causes.

Positive identification and electronic access control, governed by IEEE Standard P1402[21], currently under revision, is now the established, minimum requirement for security at power substations [18]. It describes the requirements for positive access control, monitoring of facilities, and delay/deter features.

The new systems to detect intruders and environmental hazards at electric transmission towers and substations are using a new generation of distributed

20. Video analytics, also referred to as video content analysis (VCA), is a generic term used to describe computerized processing and analysis of video streams and is typically associated with analysis of video streams captured by surveillance systems.

21. The convergence of physical and cyber security means that IEEE Standard P1402 also is referenced in cyber security requirements. In fact, virtually all cyber security standards reference physical security. Formerly, IEEE P1402 was a "guideline" but, due to increased intrusion, theft, and the threat of terrorism, it is being upgraded to a standard. According to this standard, the degree of monitoring and access control must be designed to match the nature of the threat.

sensor networks based on advanced wireless mesh networking technology. They use different sensors for different applications; the geophone/magnetometer sensor nodes are designed to be buried in the ground and to detect ground motion from pedestrians and vehicles, and they detect changes to the magnetic field caused by the passage of vehicles.

The passive infrared/accelerometer/thermistor sensor nodes are intended for mounting above ground. They detect the motion of intruders and vibrations on perimeter fences, transmission towers, and transformers resulting from intrusion or tampering, and they sense temperatures and differential temperatures for detection of extreme environmental conditions, including wildfire and transformer's state of health [19].

5.6 Review Questions and Problems

1. Describe and compare the most important parameters of the communications channel used for SCADA and teleprotection applications.
2. Is there a specific frequency band/subband used only for power utility applications? Discuss the topic.
3. Why are optical fibers are preferred transmission media for power utilities?
4. What is a difference between critical and noncritical traffic when it comes to power utility applications? What are the requirements for reliability and availability of these two systems?
5. The number of regenerators on a 200-mile-long fiber-optic system will be three (3) in addition to two (2) terminal multiplexers. Assume the worst-case scenario where the maximum transmission latency for any of the STS-1's cross-connected from one OC-48 main to the other (pass-through) is 80 μsec, and the maximum transmission latency for any of the VT1.5's (DS-1) dropped from one OC-48 main to a DS-1 port on a DS-1 card is 180 μsec. The assumption is that all the regenerators will just pass the signal, while the signal will be dropped to DS-1 level only at the terminal ends.
   A. Calculate a total end-to-end latency of the 200-mile-long OPGW system.

$$\begin{aligned}\text{Fiber-Optic Propagation Delay} &= 200 \text{ miles} \times 0.008 \text{ ms/mile} = 1.6 \text{ ms}\\ \text{Fiber-Optic Equipment Delay} &= 0.18 \text{ ms} \times 2 + 0.08 \text{ ms} \times 3 = 0.6 \text{ ms}\\ \text{Total Fiber-Optic End-to-End Delay} &= 1.6 \text{ ms} + 0.6 \text{ ms} = 2.2 \text{ ms}\end{aligned}$$

   B. Repeat the calculation for a 115-mile-long fiber-optic system requiring only one regenerator. (Answer: 1.36 ms.)
6. Show mathematically that, after a time equal to the MTBF, only 37% of the 10,000 installed pieces of equipment are expected to still be working.
7. A preliminary microwave plan calls for an estimate of a one-way latency of a 6-GHz microwave long-haul system consisting of eight microwave links. Three links are less than 35 miles in length, and four links are between 40 and 50 miles in length.

A. Estimate the worst case of total end-to-end latency, including the path propagation and equipment latency.

B. How many links can we connect in series without exceeding 16 ms end-to-end latency? Explain your reasoning.

8. Create a comparison for a 200-mile-long, 100-Mbps transmission system using different transmission media. Analyze the pros and cons of using a fiber-optic system, a satellite system, and a microwave radio point-to-point system and make assumptions for the input data that is missing in this problem. Provide the results in the form of a comparison table and answer the following questions:

   A. What are the parameters that will affect the final decision of the design engineer?

   B. How are signal latency and communications system capacity impacting the decision?

   C. Where and when should engineers consider the use of satellite communications for the communications system in the power utility space? Please elaborate.

   D. First make a decision using only technical data and then include the financial information and reevaluate your previous decision. Will the decision change?

9. A federal analysis reported by The Wall Street Journal (March 12, 2014) showed that if only nine of the country's 55,000 electrical substations were to go down—whether from mechanical issues or malicious attack—the nation would be plunged into a coast-to-coast blackout. (One month later, sniper fire knocked out a substation in San Jose, California).

   Other reports, meanwhile, have said that virtually all the country's gas and electric utilities rely on vulnerable operating systems at workstations and that the electrical grid at large stands susceptible to cyberattack from abroad and organized criminal networks.

   A. Please investigate these two statements and discuss each of them separately.

   B. Why would organized criminal networks want to get involved with cyberattacks on the electric utilities?

10. American President Barack Obama launched a new sanctions program on April 1, 2015, to target individuals and groups outside the United States that use "malicious cyberattacks" to threaten U.S. foreign policy, national security, or economic stability. In an executive order, Obama declared the consequences of such activities to constitute a "national emergency" and allowed the U.S. Treasury Department to freeze the assets and bar other financial transactions of entities engaged in destructive cyber attacks.

    A. Please discuss this move by the U.S. government.

    B. What else should governments around the world do to prevent/reduce hacking attacks on businesses and critical infrastructure?

References

[1] WECC, *Communications Systems Performance Guide for Electric Protection Systems*, 2013.

[2] Lehpamer, H., *Transmission Systems Design Handbook for Wireless Networks*, Norwood, MA: Artech House, 2002.

[3] IEEE Power & Energy Society, *IEEE Recommended Practice for the Electrical Protection of Communication Facilities Serving Electric Supply Locations Using Optical Fiber Systems*, IEEE Std. 1590-2009.

[4] Dahlin, T., et al., *Pilot Protection Communication Channel Requirements*, RFL, White Paper, 2004.

[5] NERC, *Reliability Standards for the Bulk Electric Systems of North America*, 2008.

[6] WECC, *Guidelines for the Design of Critical Communications Circuits*, 2013.

[7] Alcoa Fujikura Ltd., *Reliability of Fiber Optic Cable Systems: Buried Fiber Optic Cable, Optical Groundwire Cable, All Dielectric Self Supporting Cable*, May 2001.

[8] Telcordia GR-418-CORE, *Generic Reliability Assurance Requirements for Fiber Optic Transport Systems*, 1999.

[9] Telcordia SR-332, *Reliability Prediction Procedure for Electronic Equipment (RPP)*, 2006.

[10] NERC, *Protection System Reliability—Redundancy of Protection System Elements*, Technical Paper, 2008.

[11] IEEE Std C37.236-2013, *IEEE Guide for Power System Protective Relay Applications Over Digital Communication Channels*, 2013.

[12] Alstom, "*Network Protection and Automation-Protective Relays, Measurement & Control*," 2011.

[13] Blackburn J. L., and J. T. Domin, *Protective Relaying: Principles and Applications*, Fourth Edition, Boca Raton, FL: CRC Press, 2014.

[14] Abel, A., "Government Activities to Protect the Electric Grid," *CRS Report for Congress*, Order Code RS21958, February 4, 2005.

[15] Chen, M. T., "Petri Net Modeling of Cyber-Physical Attacks on Smart Grid," *IEEE Transactions on Smart Grid*, Vol. 2, No. 4, December 2011.

[16] http://www.ferc.gov/industries/electric/indus-act/reliability/cybersecurity.asp (accessed March 23, 2015).

[17] Krutz, R. L., *Securing SCADA Systems*, Indianapolis, IN: Wiley Publishing, Inc., 2006.

[18] IEEE Std. 1402-2000, *Guide for Electric Power Substation Physical and Electronic Security*, 2000.

[19] SAIC, *Advanced Distributed Sensor Networks for Electric Utilities*, Public Interest Energy Research (PIER) Program, Final Report, Prepared for California Energy Commission, June 2012.

CHAPTER 6

Regulatory and Safety Challenges

6.1 Licensed and License-Exempt Radio Systems

Licensed RF transmitters communicate using a specific transmit and receive frequency combination that is assigned to the user (licensee). The frequency assignment is coordinated with other users of the same spectrum in the same geographical area. This process provides full disclosure of the frequency assignment and typically avoids interference from any existing licensee in the area. If licensed radios encounter interference, it is typically resolved with the assistance of the regulatory body.

The terms *license-exempt* (also called *license-free*) and *licensed* refer to the radio frequency spectrum rules defined by the FCC. In the United States, FCC Rules Part 15 governs the license-exempt frequency spectrum, and Rules Part 101 governs the licensed frequency spectrum. Licensed products require regulatory approval before deployment, whereas license-exempt products can be deployed without any regulatory approval. A *license-exempt* system can be installed quickly virtually anywhere within a given country without obtaining an operational license from the regulatory authorities.[1]

Most countries have government agencies responsible for allocating and controlling the use of the radio spectrum. In the United States, spectrum is allocated by the FCC for commercial use and by the *Office of Spectrum Management* (OSM)[2] for military and government use. Commercial spectral allocation is governed in Europe by ETSI and globally by the ITU.

6.2 Spread-Spectrum Radio Communications

Spread-spectrum systems were originally developed by the military to counter attempts to detect, decode, or block signal transmissions. The most important

1. Such a system must already be certified to operate as license-exempt in that country. Manufacturers desiring license-exempt certification must apply to the FCC or equivalent national authority for the type approval to operate the particular product in specific radio frequency bands.

2. OSM principally manages the Federal government's use of the radio frequency spectrum. To achieve this, OSM receives assistance and advice from the Interdepartment Radio Advisory Committee (IRAC). It is part of NTIA.

nonmilitary characteristics of spread spectrum systems are that they facilitate radio communications in a manner that minimizes the potential to cause harmful interference to other services and that they are able to withstand higher levels of interference than other technologies. Hence, spread-spectrum systems have significant potential to share common spectrum with other services.

Spread-spectrum devices operate on a license-exempt basis[3] if the technical conditions are met, and type approval is mandatory. Two main types of spread-spectrum systems are commercially available: *direct sequence* (DS) and *frequency hopping* (FH).

DS is where the incoming information is usually digitized (if not already in binary format) and then "modulo 2" is added to a higher-speed code sequence. The combined information and the code are then used to modulate a radio frequency carrier. Since the high-speed code sequence dominates the modulating function, it is the direct cause of the wide spreading of the transmitted signal.

In *FH spread spectrum* (FHSS) the carrier is modulated with coded information in a conventional manner, causing a conventional spreading of the radio frequency energy about the carrier frequency. The frequency of the carrier is not fixed but changes rapidly under the direction of a pseudorandom coded sequence. The wide radio frequency bandwidth needed by such a system is not required to support a spreading of the radio frequency energy about the carrier, but rather to accommodate the range of frequencies to which the carrier frequency can hop.

The frequencies used for the operation of spread-spectrum devices have been mainly in bands designated for ISM applications at 900 MHz, 2.4 GHz[4], and (to a lesser extent) at 5.8 GHz.

In part for reasons of safety, the transmitter power output level in the ISM band is limited to 1W (+30 dBm) maximum input antenna power. For similar reasons, and to minimize interference, EIRP, or power radiated by the associated antenna system, is limited to 4W (36 dBm) maximum.

In the United States (2.4-GHz band), for every 3 dB of antenna gain over +6 dBi, the input power to the antenna is reduced by 1 dB from +30 dBm. For now, in the 5.8-GHz band, there are no EIRP limits—only a +30 dBm maximum antenna input power.[5]

The 5-GHz band is divided into several sections referred to as U-NII bands. (See Table 6.1.) The FCC is currently considering the addition of two more U-NII bands in the 5-GHz band; this would be U-NII-2B, 5.350–5.470 GHz and U-NII-4, 5.850 – 5.925 GHz.

Dynamic frequency selection (DFS) is a mechanism to allow license-exempt devices to share spectrum with existing military radar and weather radar systems. At power-up and throughout operation, each point-to-point terminal scans the band

3. It is important to note that not all the license-exempt microwave systems are spread-spectrum.

4. The 2.4-GHz ISM band used to be called the "junk band," because it was already contaminated by microwave oven emissions. Years ago, 2.43 GHz was allocated to the microwave oven, and it was felt that no one would ever want to co-occupy this band. As pressure to allocate more spectrum to communications was felt, the FCC set up rules for license-exempt ISM operation in this band.

5. This might change soon, since FCC is considering some additional power limitations in 5.8 GHz license-exempt frequency band.

Table 6.1 U-NII Bands

Band	*Frequency Range [MHz]*	*Number of Channels*	*Max Power*	*Max EIRP*
U-NII-1(Indoor/Outdoor)	5,150–5,250	4	250 mW client 1W master	1-W client 4-W master 200-W PTP
U-NII 2A (Indoor/outdoor)	5,250–5,350	4	250 mW	1W (30 dBm)
U-NII 2C (Worldwide, indoor/outdoor)	5,470–5,725	11	250 mW	1W (30 dBm)
U-NII 3 (Indoor/outdoor)	5,725–5,850	4	1W	4W No limit for PTP

hundreds of times per second and automatically switches to the clearest channel.[6] Some radios will have the 30-day, time-stamped database to alert network operators to any interference that does exist and provide statistics that help pinpoint which channels offer the clearest data paths.

In addition, all U-NII devices must contain security features to protect against modification of software by unauthorized parties.

Regulations in the license-exempt bands keep changing so it is recommended to check for the latest updates.[7]

6.3 Traffic Safety and FAA Regulations

Under authority granted by the Federal Aviation Act, the FAA has jurisdiction over the following communication facilities: towers that exceed 200 ft in height, towers that are located within 20,000 ft of a major commercial or military airport, or towers that are located within 10,000 ft of a general aviation airport.

The FAA reviews the location and height of such towers and may require them to be painted and/or illuminated to prevent possible interference with nearby airport operations. The FAA also reviews possible interference issues with aircraft-to-ground communications that may be caused by transmission facilities located in or near airport flight paths.

In the United States, owners of antenna towers that are taller than 200 ft (60m) above the ground level or that may intersect the flight pathways of a nearby airport, must register the structure with the FCC. In addition, they need to have the

6. The process for a radio to detect and identify a radar signal is a complicated task that sometimes leads to incorrect results due to a large number of factors, including uncertainties of the RF environment and the ability of the access point to reliably detect actual on channel radar.

7. Some licensed users sometimes operate in the license-exempt bands as well (for example, U.S. government radar equipment in the U.S. U-NII band at 5.725 to 5.825 GHz). These radars often operate at peak power levels of millions of watts, which can cause significant interference problems to other nearby users in this band. Therefore, it is important to determine if there are any airports, military bases, or other installations where such radars may be located.

structure's placement studied by the FAA.[8] Once the FAA has completed an aeronautical study, a determination is made regarding the impact to air navigation.

For air navigation, where precision is important, it is vital to minimize the chances of error or system outage. Much like other RF systems, navigational aids are also subject to interference that may negatively impact the system, causing errors and potentially disabling the system entirely. The interference effects of interest are shadowing, reflections, and Fresnel zone clearance.

There are a number of different navigational aids, but we will briefly discuss only VORTAC[9], a navigational aid for aircraft pilots. It consists of a colocated *VHF omni-directional range* (VOR) *beacon* and a *tactical air navigation system* (TACAN) beacon. Both types of beacons provide pilots azimuth information, but the VOR system is generally used by civil aircrafts and the TACAN system by military aircrafts. However, the TACAN distance-measuring equipment is also used for civil applications because it is built to civil specifications.

Typically, it is the best to stay outside of 2 nautical miles (NM) from the VORTAC site when erecting large structures (e.g., transmission towers, electric lines, communication towers, and wind turbines) [1].

6.4 Human Exposure to RF Electromagnetic Fields

People are exposed to different types of radiation every day; in the United States, the FCC is required by the National Environmental Policy Act (NEPA) of 1969, to, among other things, evaluate the effect of emissions from FCC-regulated transmitters on the quality of the human environment.

Several organizations, such as the ANSI, the IEEE, and the *National Council on Radiation Protection and Measurements* (NCRP) have issued recommendations for human exposure to RF electromagnetic fields. On August 1, 1996, the Commission adopted the NCRP's recommended *maximum permissible exposure* (MPE)[10] limits for field strength and power density for the transmitters operating at frequencies of 300 kHz–100 GHz. In addition, the Commission adopted the *specific absorption rate* (SAR)[11] limits for devices operating within close proximity to the body as specified within the ANSI/IEEE C95.1-1992 guidelines.

8. FAA guidelines regarding towers come from its Electronic Code of Federal Regulation, Title 14, Part 77, www.tinyurl.com/faa-fedreg. It establishes standards and notification requirements for objects affecting navigable airspace.

9. VORTACs are old, ground-based navigational technology (regulated by FAA Order 6820.10, published in 1985), and many of them will be out of service very soon or replaced with satellite-based navigational systems. In the meantime, we still have to pay close attention to VORTAC site locations. VOR and TACAN locations are public information and available on the Internet.

10. Maximum permissible exposure (MPE) is the rms and peak electric and magnetic field strength, their squares, or the plane-wave equivalent power densities associated with these fields to which a person may be exposed without harmful effect and with an acceptable safety factor.

11. SAR is a measure of the rate of energy absorbed by (dissipated in) in the dielectric materials such as biological tissues. SAR is usually expressed in terms of watts per kilogram (W/kg) or milliwatts per gram (mW/g). Guidelines for human exposure to RF fields are based on SAR thresholds where adverse biological effects may occur. When the human body is exposed to an RF field, the SAR experienced is proportional to the squared value of the electric field strength induced in the body.

In general, the RF EMF exposure assessment of the *general public* and *workers (occupational exposure)* use the same methods. However, there are some specific features that distinguish the two categories. The most important are listed as follows:

- Exposure limits for the general public are more conservative because people have less knowledge about possible hazards and location of places with the highest exposure levels.
- Workers are more likely to more closely approach RF sources (transmitting antennas) so in more cases the area under examination is located in the reactive near-field region, which, as a consequence, requires more sophisticated methods for the exposure assessment by measurement or by calculation.
- Workers in some cases are allowed to stay in areas with exposure levels above the limits but for limited time periods or by using mitigation measures like protective clothes.

If the area under consideration is covered by emissions from existing RF sources only, then the exposure assessment can be performed either by measurements or by calculations. If in the area under consideration some new RF EMF sources are planned (even if there exist other operating radio systems), the possibility to check compliance is through the exposure assessment by calculation or by a combination of measurement and calculation [2].

There is a large number of parameters involved in determination of the level of human exposure to RF electromagnetic fields, including transmitter power, distance from the antenna, antenna radiation pattern, and location of the antenna(s) and their surroundings. Figure 6.1 shows plane-wave equivalent power density limits for general public and occupational exposure as a function of frequency. Simplified analysis and limits of exposure are described in FCC OET Bulletin 65 [3] and in IEEE Std. C95.1-2005.

6.5 Telecommunication Circuits Interfacing High-Voltage Environment

Telecommunications service to electric power stations is provided via a link that connects the telecommunications equipment located at or near the electric power station with the CO or switching center of the telecommunications service provider. The telecommunications facility may be a metallic wire-line cable, an optical fiber cable, radio system, or a microwave link.

Wire-line telecommunication facilities serving electric supply locations often require special *high-voltage protection* (HVP) against the effects of fault-produced *ground potential rise* (GPR) or induced voltages, or both. Effective protection of any wire-line telecommunication circuit requires coordinated protection on all circuits provided over the same telecommunication cable.

6.5.1 Grounding and GPR

Proper grounding is a very important factor in protecting human lives and expensive equipment that can be found inside of the substation perimeter. The performance of

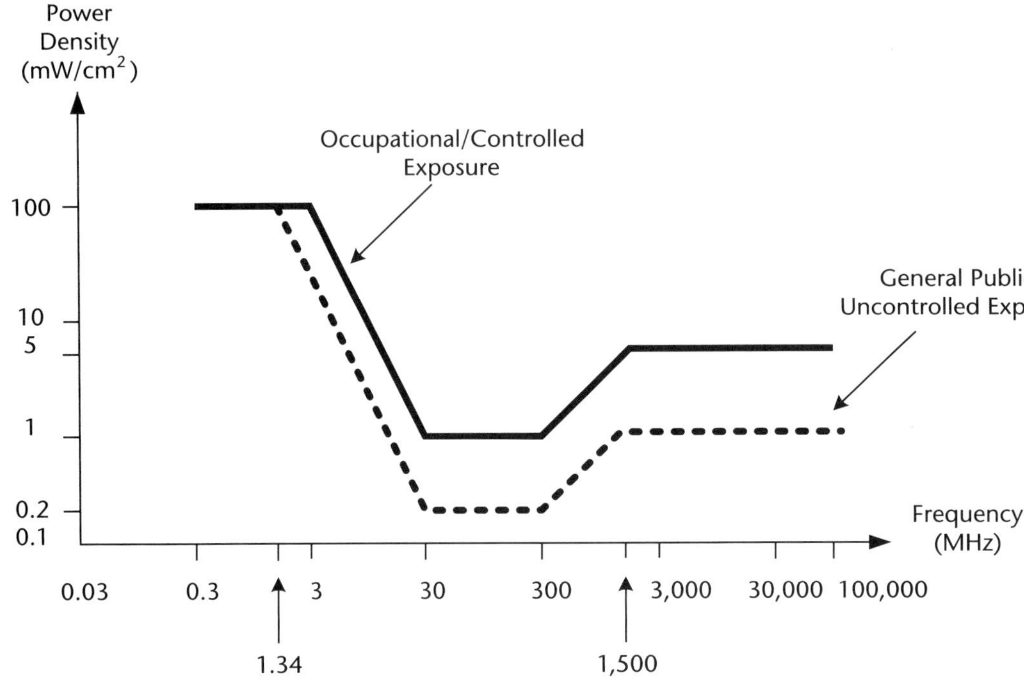

Figure 6.1 FCC limits for MPE.

a grounding grid is affected by a number of factors—soil conductivity and grounding rod corrosion, for example.

When a power system ground fault (caused by lightning, for example) occurs, all or some of the fault current returns via the Earth through the ground grid and produces a potential difference between the ground grid and remote Earth. This potential difference is defined as power station GPR. The *voltage profile*[12] of the GPR is shown in Figure 6.2.

The resultant voltage not only causes a potential difference between the station ground and the true Earth, but it also affects the surrounding Earth outside the station. In case a substation has a metallic fence, it must be properly grounded to protect people from high voltages that may occur during a fault in the network.

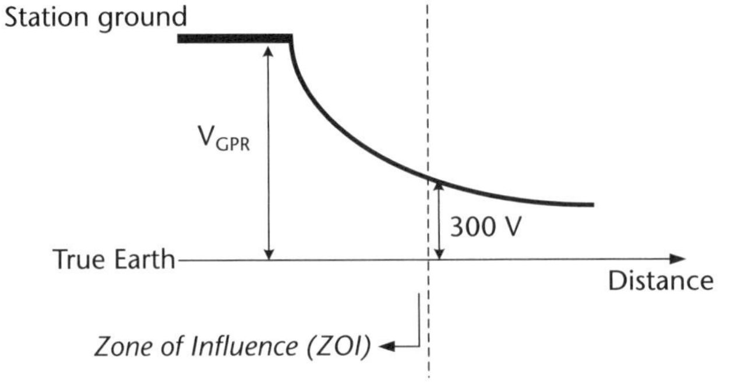

Figure 6.2 Station GPR.

12. The grounding study will determine the 300-V line. It is the distance from the fault location to the point where the GPR value has diminished to 300V, which can be from a few feet to thousands of feet.

Currents flowing in the Earth's surface during a fault can cause metal objects to have a significantly different voltage than the ground under a person's feet; this touch potential presents a hazard of electrocution. Any other conducting object connected to the substation Earth ground, such as telephone wires, rails, or metallic piping, may also be energized at the ground potential in the substation. This effect will decrease as distance from the substation increases.

The amount of GPR is a function of the amount of grounding electrodes in contact with the soil, soil resistivity, and various other factors, such as the number of conductors leaving the site [4]. Soil resistivity is a function of the concentrations of ionic soluble salts and the moisture level in the soil, and testing needs to be done to measure how well or how poorly the soil will conduct electric current. Soil resistivity testing is absolutely essential when it comes to reliable electrical grounding design or determining ground rise potential.

The grounding electrode resistance of large electrical substations should be 1 ohm or less. For commercial and industrial substations, including telecommunications sites, the recommended resistance to ground is 5 ohms or less.

6.5.2 GPR and Telecom Circuits

If equipment is all tied to the same ground grid and is not referenced to any external ground, then it will not be damaged due to GPR. In case of leased lines from local Telco, communications circuits are connected through equipment bonded to the substation's ground grid but also terminated to a CO by a copper wire (Figure 6.3).

This CO is the remote Earth, and the copper wire-line is a conductor tied between two ground planes. Therefore, a current will tend to flow, which is dangerous to personnel and can damage the site and its equipment. As the *channel service unit* (CSU) is usually not hardened for use within a substation, nearby faults can cause the channel to be lost. A detailed GPR study is required for every site that has connections with the remote sites that have their own grounding system. GPR studies are performed when telecommunications sites (or just circuits) are to be installed in the high-voltage environments found in power distribution systems and substations. When leasing T1 circuits from the local telephone company, GPR

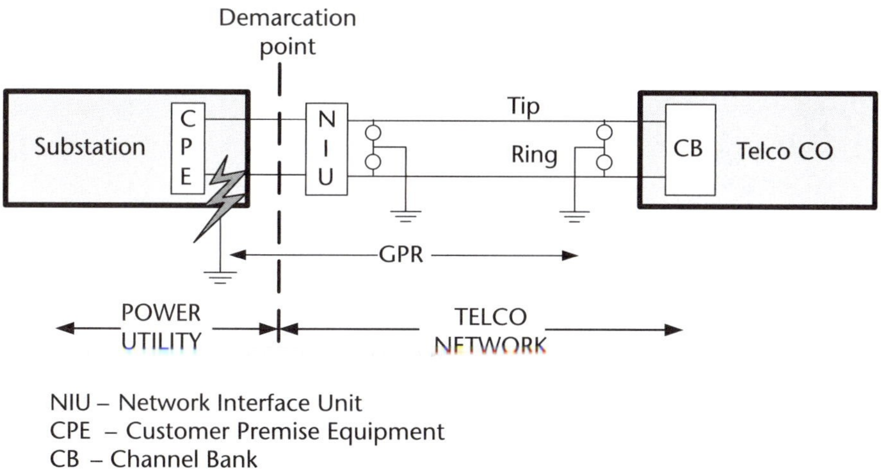

Figure 6.3 GPR and leased lines.

study calculations are made according to IEEE Std. 367 [5] and will be supplied to the telco. The telco will dictate its interface requirements based on its own standard procedures.[13]

The output of the study is the 300-V line, recommendations to make the site safe from a *step and touch potential*, and the recommendations for the grounding system that will satisfy the telecom carrier's requirement.

The *network interface unit* (NIU) will hand off the circuit to a *customer premise equipment* (CPE) device, most likely a channel bank. It is important to ensure that the CO side of the T1 circuit is isolated from the utility grounding system, according to IEEE standards [6].

The three categories of *service performance objectives* (SPOs), as defined in IEEE Standard 487-2007, are described as follows:

- Class A: A noninterruptible service performance that operates before, during, and after a power fault condition. Class A requires special design and circuit coordination between the power utility and telco. To fully achieve class A the use of dual alternate routing should be considered. Examples are pilot-wire protective relaying, audio-tone protective relaying, and critical supervisory circuits that are used for remote control of a power network.
- Class B: An interruptible service performance that is self-restoring and must operate before and after a power fault condition, but not necessarily during a power fault condition. Class B requires special design and circuit coordination between power utility and telco. Examples are emergency telephone, high-priority data collection, supervisory control, and signal and alarm circuits that would be required to put a power network back into operation after a power fault.
- Class C: An interruptible service performance that can tolerate a station visit to restore service. Examples are POTS, noncritical telemetering and data, and some signal and alarm circuits that would not be used for network restoration.

6.5.3 Equipment and Methods to Resolve GPR

Wire-line telecommunications facilities, serving electric supply locations, often require special high-voltage protection against the effects of fault-produced GPR or induced voltages, or both [7].

There are three items from a safety point of view that must be considered when placing communications facilities within high-voltage environments:

- Step voltage: The difference in surface potential experienced by a person bridging a distance of 1m with their feet without contacting any grounded objects.
- Touch voltage: The potential difference between the GPR and the surface potential at the point where a person is standing while at the same time having a hand in contact with a grounded structure.

13. Special high-voltage protection requirements are reviewed by the local telco's protection department. The time required for this type of service is typically a minimum of three to four months.

- Transferred voltage: The special case of the touch voltage where a voltage is transferred into or out of the substation from, or to, a remote point external to the substation site.

IEEE Std. 487 states that high-voltage protection is required at sites with a peak GPR greater than 1,000V. For sites with a GPR of 1,000Vpeak or less, gas tubes or other shunting devices are suitable. However, a gas-tube operation will cut power for the span and the digital signal (during that period, the network would be unavailable).

Figure 6.4 depicts the principle of isolation of remote circuits entering a substation environment. Transition from regular copper wiring to fiber-optic cable is preferably performed outside of *zone of influence* (ZOI), and all-dielectric fiber-optic cable is then brought into the substation [8]. This fiber-optic cable, now inside of the electric utility site, can be connected directly to the CPE or converted back to copper interface (as shown here) and then connected to the CPE.

The *copper-fiber junction* (CFJ) is typically powered by a remote ground referenced, locally derived or isolated source such as a CO line power, a locally powered ac-to-dc UPS system, power-over-fiber (PoF) powering, or solar power.

The U.S. Occupational Safety and Health Administration (OSHA) has designated GPR as a "known hazard" and has issued regulations governing the elimination of this hazard in the work place. Protection and isolation equipment is made to national and international standards described by IEEE, FCC, National Electrical Codes (UL/CSA), and Telcordia.

6.6 Communications Equipment in Substations

Every substation is a part of an electrical generation, transmission, and distribution system. Substations transform voltage from high to low, or the reverse, and perform other important functions. Substations may be owned and operated by an electrical

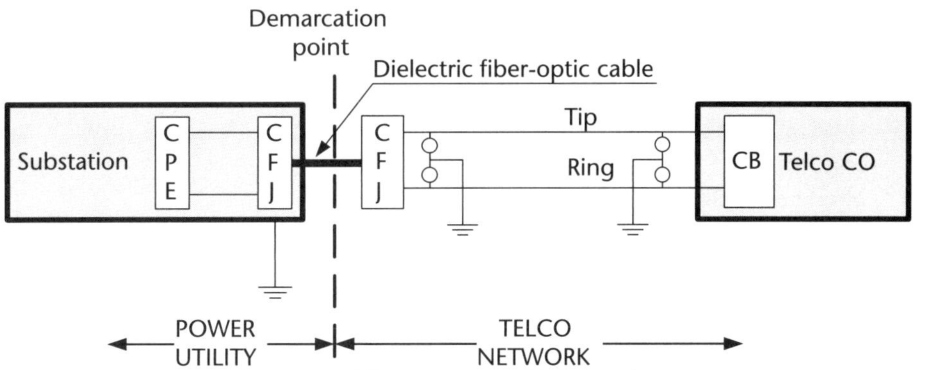

CPE–Customer Premise Equipment
CB–Channel Bank
CFJ–Copper-Fiber Junction

Figure 6.4 Electrical isolation of remote circuits.

utility, or may be owned by a large industrial or commercial customer. Generally, substations are unattended, relying on SCADA for remote supervision and control.

Substations generally have transformers and switching, protection, and control equipment. In a large substation, circuit breakers are used to interrupt any short circuits or overload currents that may occur on the network. Smaller distribution stations may use recloser circuit breakers or fuses for protection of distribution circuits. Substations themselves do not usually have generators, although a power plant may have a substation nearby. Other devices such as capacitors and voltage regulators may also be located at a substation.

Substations may be on the surface in fenced enclosures, underground, or located in special-purpose buildings. High-rise buildings may have several indoor substations. Indoor substations are usually found in urban areas to reduce the noise from the transformers, for reasons of appearance, or to protect switchgear from extreme climate or pollution conditions.

6.6.1 Substation Automation Structure

Modern substation automation is structured in three basic levels:

- The station level, which provides an overview across the whole station, is located in a shielded control room. The station level includes equipment such as the HMI work station, the master station computer, the backup station computer, and the GPS receiver.
- The bay level conducts maintenance work only within one bay, and it is usually close to the switchgear. The bay level includes protection and control IEDs of different bays such as circuit breakers, transformers, and capacitor banks. Equipment in bay level and station level is called secondary equipment.
- The process level provides the interface between the substation automation system and the switchgear. The process level includes switchyard equipment (also primary equipment) such as CTs/PTs, remote I/O, actuators, and merging units.

In order to sustain reliable communication and to avoid catastrophic failures, power substations require substation-hardened communications systems against environmental influences of mechanical, thermal, EMC, or other harsh operating conditions. This is especially important when it comes to relay protection and related communication equipment.

6.6.2 EMI

Substation equipment such as transformers, bays, and other electrical devices will create electromagnetic fields. Electrostatic discharge, vibration, and high-energy power surges will occur during the substation operation. Two classes of EMI immunity for communication devices are defined as follows:

- In class 1, devices may experience some data errors, loss, or delays under EMI stress conditions, which is adequate for network supervision.
- In class 2, devices must not experience any data errors, delays, or loss under EMI stress conditions, which is essential for IED-to-IED communication in substations.

6.6.3 Environmental Standards Compliance

The communication of devices inside the substation environment might be damaged or destroyed by environmental disturbances. The most critical environmental conditions in a substation are EMC and climatic conditions. Thus, as mission-critical components, extremely robust, ruggedized communications devices are required to ensure correct operation.

The general environment requirements for industrial applications are not considered sufficient for substations. Communications devices in substations must comply with the environmental conditions described in the following standards:

- IEEE 1613 presents IEEE standard environmental and testing requirements for communications networking devices in electric power substations. Communications devices can be specified here in terms of such factors as temperature and humidity, power supply, high-voltage resistance, EMC and radiated immunity, vibration and shock, and altitude.
- IEC 61850-3 is the main standard for substation environmental conditions. It lists the environmental requirements for factors such as supply voltage, EMI immunity, temperature, humidity, barometric pressure, mechanical, seismic, pollution, and corrosion. Other IEC standards, like IEC 60870-2 (environmental climatic conditions), are referred in IEC 61850-3.

6.6.4 Antenna and Radio Installation Safety and Security Issues

Tower construction and radio/microwave antenna installation require special skills. (See Figure 6.5.) Without the cranes to lower pieces into place from above, the towers are constructed with an apparatus known as a gin pole. The gin pole is used to hoist material from the ground and past the worker at the top so that it can be positioned and bolted in place. The work is carried out at the mercy of wind and inclement weather at heights that reach up to a few hundred meters. Typically, communications towers range between 30 and 150m (100–500 ft) in height, while broadcast towers can be much higher.

It is obvious that communications tower safety is a life-and-death issue, not only for the men and women who work on the structures, but also for the companies that own them. While working on a communications tower, very stringent work procedures and reliable safety equipment are required to protect workers from falls.

Figure 6.5 Microwave antenna installation in the substation environment.

Under OSHA, employers are responsible for providing a safe and healthful workplace.[14]

Prior to the 1980s, communications and broadcast tower erection, servicing, and maintenance were a very small and highly specialized industry; the growing demand for radio and broadcast communications has spurred a huge increase in communications tower construction and maintenance. Some of the more frequently encountered risks include falls from great heights, electrical hazards, hazards associated with hoisting personnel and equipment with base-mounted drum hoists, inclement weather, falling object hazards, equipment failure, and structural collapse of towers.

Despite the common practice of contracting out the construction and maintenance of the towers to other companies, the owners of the structure (company directors, officers, and shareholders) share a legal responsibility with the subcontractor to make sure the work is carried out safely. This situation is not a new one—most jurisdictions require property owners to provide safe job sites regardless of whether they directly employ the workers.

In addition, a contractor must carry sufficient liability insurance for services provided within (or outside) North America, and the copy of the insurance cover-

14. With the Occupational Safety and Health Act of 1970, Congress created the Occupational Safety and Health Administration (OSHA) to assure safe and healthful working conditions for working men and women by setting and enforcing standards and by providing training, outreach, education and assistance.

age document has to be provided to the project manager prior to the start of any installation.

6.7 Review Questions and Problems

1. During the 1920s, anyone possessing radio equipment was allowed to broadcast signals over the air, resulting in chaos. In 1934, Congress created the FCC to regulate radio communications in the United States, the District of Columbia, and all U.S. possessions [9]. The FCC has historically controlled access to radio spectrum by allocating specific frequency bands for use by licensed service providers.
   A. Today, many applications take advantage of the so-called license-exempt frequency bands. Identify the spectrum bands that can be used without a license in your country.
   B. Research different applications and describe advantages and disadvantages of wide use of these frequency bands. Show your results in the form of a table and briefly summarize your findings.
2. What are the network security requirements of an electric power utility? Are oil/gas/water utilities security requirements any different from electric utilities? What about the physical security? Please compare and elaborate.
3. The utility would like to use license-exempt 900-MHz radio for its SCADA application. It chooses 11-dBi Yagi-Uda antennas and needs 30 ft of coaxial (4 dB/100 ft) cable between the radio and the antenna. Assuming that total connectors and surge protector losses is 1 dB, what is the maximum allowed transmitter power of the radio to stay in compliance with FCC? (Hint: Calculate EIRP. Answer: 27.2 dBm.)
4. What are the main parameters of interest when licensing a new radio system? Describe the process of designing a new microwave link and path engineering process.
5. Are fiber-optic multiplexers, microwave radios, and Ethernet switches and routers available as substation-hardened? List some examples of nonhardened and hardened equipment, models, and manufacturers and specify standard compliance.
6. Discuss the importance of a grounding study when bringing the external communications circuits into the electric substation. List standards that apply to grounding of communications facilities (e.g., telco circuits, communication towers, and equipment racks).
7. During construction of a new transmission line, workers and their management were having health and safety concerns regarding small 2-ft microwave radio antennas pointing in their direction from the nearby communications tower. Please discuss their concerns. What would be your course of action in this case?
8. Discuss Question 7 from the microwave link owner's perspective; the owner's fear is that all the transmission line construction activity—cranes, lots of metal parts, and people—could block the Fresnel zone and disrupt its

communications system. Are the microwave link owner's fears justified? What would be the correct course of action here? Please discuss.

9. The National Association of Tower Erectors (NATE) is a nonprofit trade association providing a unified voice for tower erection, maintenance, and service companies [10].
   A. Research the role of OSHA and NATE in ensuring safe working conditions for tower personnel.
   B. What are the most important things to consider when hiring a contractor to install antennas on a communications tower?
10. In 2013, OSHA recorded a total number of 13 communications tower-related fatalities, and it seems that there is a significant increase in fatalities and injuries every year. OSHA is working with industry stakeholders to identify the causes of these injuries and fatalities and to reduce the risks faced by employees in the communications tower industry. Research and discuss the increase in fatalities related to communications towers.

References

[1] FAA Order 6820.10, VOR VOR/DME, and VORTAC Siting Criteria, 1985.

[2] ITU-T, Recommendation K.91-2012, *Guidance for Assessment, Evaluation and Monitoring of Human Exposure to Radio Frequency Electromagnetic Fields*, 2012.

[3] FCC, Office of Engineering & Technology, OET Bulletin 65, Edition 97-01, *Evaluating Compliance with FCC Guidelines for Human Exposure to Radiofrequency Electromagnetic Fields*, 1997.

[4] IEEE Std 81, *IEEE Guide for Measuring Earth Resistivity, Ground Impedance, and Earth Surface Potentials of a Grounding System, 2012.*

[5] IEEE Std 367, *Recommended Practice for Determining the Electric Power Station Ground Potential Rise and Induced Voltage From a Power Fault, 2012.*

[6] IEEE Std 487, *Recommended Practice for the Protection of Wire-Line Communication Facilities Serving Electric Supply Locations, 2007.*

[7] IEEE Std 80, *IEEE Guide for Safety in AC Substation Grounding, 2000.*

[8] IEEE Std. 487.3, *Standard for the Electrical Protection of Communication Facilities Serving Electric Supply Locations Through the Use of Hybrid Facilities*, 2013.

[9] Carter, K. R., et al., "Unlicensed and Unshackled: A Joint OSP-OET White Paper on License-exempt Devices and Their Regulatory Issues," FCC, Office of Strategic Planning and Policy Analysis, OSP Working Paper Series, May 2003.

[10] http://www.natehome.com (accessed August 31, 2015).

CHAPTER 7

Installation, Testing, and Commissioning

7.1 Field Surveys and GIS Data

Every detailed design of a communications system should start with the site visit, regardless of whether the communications infrastructure is new, an upgrade of the existing infrastructure, or a complete greenfield site without any existing structures.

A *site survey report* will be generated, containing detail information about equipment room, shelter, or cabinet, with digital pictures, sketches, maps, and documented comments and observations from the field. A field survey will determine exact location (coordinates) of the site, site access, contact names, communications tower location, or fiber-optic regenerator along the electric transmission line. A surveyor will take note of surrounding natural (hills and trees) and man-made structures (e.g., electric towers and buildings) and estimate or measure their heights [1].

Drawings of the existing facilities and equipment will be obtained from the records before coming to the site or created while at the site. Existing drawings and other engineering documentation should be compared with the status in the field and updated if necessary.

7.1.1 Maps, Datums, and Geometric Earth Models

There are three general types of maps: planimetric, topographic, and orthophoto. Planimetric maps show the positions of features without showing their relationship to the hills and valleys of the land. Examples of features on planimetric maps include rivers, lakes, roads, and boundaries. Topographic maps are different from planimetric maps because they show both the horizontal and vertical (relief) positions of features. An orthophoto map is an aerial color-enhanced photograph of the land depicting terrain and other features. Some orthophoto maps are overlain with contour lines and other features commonly associated with topographic maps.

Every map is associated with a *projection* and a *spheroid*; this is called a *geodetic datum.*[1] When producing maps, the surface of the Earth has to be mapped onto a plane. This is called a *projection*. A projection from a curved surface like the

1. *Datum* from Latin, singular for data-given things. Hundreds of different datums have been used to frame position descriptions since the first estimates of the Earth's size were made by Aristotle. Referencing geodetic coordinates to the wrong datum can result in position errors of hundreds of meters.

Earth to a plane cannot be done without distortion; therefore, various projection types with different advantages and disadvantages have been developed and are still in use throughout the world.

A *spheroid* is a surface obtained by rotating an ellipse around one of its axes. In geography, the shape of the Earth is considered to be a spheroid. However, the Earth is not a perfect spheroid; therefore, several spheroid models have been adopted to approximate the surface in different parts of the Earth.

Different nations and agencies use different datums as the basis for coordinate systems used to identify positions in geographic information systems, precise positioning systems, and navigation systems. In North America, the *North American Datum 1983* (NAD83) is widely used, although the older *North American Datum 1927* (NAD27) is sometimes still used. NADCON is the U.S. Federal standard for NAD27 to NAD83 datum transformations. The GPS is based on the *World Geodetic System 1984* (WGS 84).

7.1.2 Magnetic and True North

Magnetic compass does not normally point to true north; over most of the Earth, it points at some angle east or west of true (geographic) north. The direction in which the compass needle points is referred to as magnetic north, and the angle between magnetic north and the true north direction is called magnetic declination. The terms variation, magnetic variation, and compass variation are sometimes used in place of magnetic declination.[2]

The current year's declination is oftentimes marked in degrees on the map, which is one reason it is important to have a current map. Magnetic field models are used to calculate magnetic declination by means of computer programs and can be found on the Internet.

In order to change magnetic to true headings, we will add east declination and subtract west declination. In order to convert from true to magnetic headings, we will subtract east and add west declination.

7.1.3 Coordinate Systems and Altitude

Coordinate systems to specify locations on the surface of the Earth have been used for centuries. In western geodesy the equator, the Tropics of Cancer and Capricorn, and lines of latitude and longitude were used to locate positions on the Earth. Eastern cartographers used other rectangular grid systems as early as 270 CE[3], and various units of length and angular distance have been used over history.

The most commonly used coordinate system today is the *latitude, longitude, and altitude system*. The prime meridian and the equator are the reference planes

2. The magnetic declination does not remain constant in time. Complex fluid motion in the outer core of the Earth (the molten metallic region that lies from 2,800 to 5,000 km below the Earth's surface) causes the magnetic field to change slowly with time. This change is known to as secular variation.

3. Common/Christian Era, or A.D. 270, Anno Domini.

used to define latitude and longitude. The geographical coordinates are typically expressed in the *sexagesimal* form [base is 60 (i.e., degrees, minutes, and seconds)].[4]

Latitude measures how far north or south of the equator a place is located. Longitude measures how far east or west of the prime meridian a place is located. The prime meridian runs through Greenwich, England. Longitude is measured in terms of east, implied by a positive number, or west, implied by a negative number.

Altitude is measured with reference to *mean sea level* (MSL), and height is measured with reference to the soil or ground level. In this application, altitudes are known as *above mean sea level* (AMSL) and heights as *above ground level* (AGL). It is essential to show the altimetric or *vertical reference datum* (VRD) with the altitude, because altitudes change when the vertical datum changes.

Sometimes the coordinates can be shown in *Universal Transverse Mercator* (UTM)[5] format or in values expressing north or east, for example. When the positioning coordinates are obtained from topographical charts, they are normally expressed in UTM format. UTM is a global coordinate system that is defined in meters rather than degrees-minutes-seconds. UTM is a very precise method of defining geographic locations; therefore, it is commonly used in GPS and GIS mapping.

The UTM grid divides the world into 60 north-south zones; the zones are numbered 1–60, and each zone is six-degrees-wide in longitude. For example, the U.S. mainland goes from Zone 10 to 19. UTM uses two coordinates, Easting and Northing, to determine a location. Locations within a zone are measured in meters east and west from the central meridian, and north and south from the equator. Coordinates include the zone, easting coordinate, and northing coordinate; this coordinate describes a specific location using meters. When using the UTM coordinate system, a location can be identified within a meter (3 feet) [2].

The *state plane coordinate system* (SPS or SPCS)[6] is a coordinate system that divides the 50 states of the United States, Puerto Rico, and the U.S. Virgin Islands into more than 120 numbered sections, referred to as *zones*. The boundaries of these zones always follow state lines and usually follow county lines. SPCS uses a simple Cartesian coordinate system to specify locations rather than a more complex spherical coordinate system (i.e., latitude and longitude). By ignoring the curvature of the Earth, simplified "plane surveying" methods can be used, while remaining highly accurate within each zone (error less than 1:10,000). Outside specific state plane zones, accuracy rapidly declines; thus the system is not useful for regional or national mapping.[7]

4. Sexagesimal is a numeral system with 60 as the base; it originated with the ancient Sumerians in the 2000s BCE, was transmitted to the Babylonians, and is still used, in modified form, for measuring time, angles, and geographic coordinates

5. The Universal Transverse Mercator coordinate system was developed by the United States Army Corps of Engineers in the 1940s and named after Gerardus Mercator, a Flemish cartographer who lived 1512–1594. The system was based on an ellipsoidal model of the Earth

6. SPCS was developed in the 1930s by the U.S. Coast and Geodetic Survey to provide a common reference system for surveyors. Officially, SPCS zones are identified by their National Geodetic Survey (NGS) code.

7. The state plane coordinate system uses a different coordinate system in each zone; in most cases this is not a problem, unless an area of interest crosses a State Plane zone boundary. Interactive maps displaying projection parameters of every system in every state can be found on Internet.

Three conformal projections were chosen to be used with this coordinate system—the *Lambert Conformal Conic* for states that are longer east–west, such as Tennessee and Kentucky; the *Transverse Mercator* projection for states that are longer north–south, such as Illinois and Vermont; and the *Oblique Mercator* projection for the panhandle of Alaska, whose maximum dimension is on a diagonal.

Originally, this coordinate system was referred to as the *State Plane Coordinate System of 1927 (SPCS 27)* and based on a network of geodetic control points referred to as the NAD27. Later, the redefined and updated system was based on NAD83[8] and called the *State Plane Coordinate System of 1983 (SPCS 83)*. The standard unit of measurement for SPCS 27 is the U.S. survey foot, while for SPCS 83, the most common unit of measure is the meter.

An older *Bureau of Land Management* (BLM) system, still used sometimes, is outdated and doesn't include codes for some of the new zones; it should therefore be avoided.

7.1.4 DEM Data

A digital elevation model (DEM), also sometimes called a digital terrain model (DTM), is a digital file consisting of terrain elevations for ground positions at regularly spaced horizontal intervals. The United States Geological Survey (USGS) produces different digital elevation products. Although all are identical in the manner the data are structured, each varies in sampling interval, geographic reference system, areas of coverage, and accuracy.

The *National Elevation Dataset* (NED) is the primary elevation data product of the USGS. The NED provides basic elevation information for Earth science studies and mapping applications in the United States and is generated at various horizontal resolutions (NED layers).

Three layers of the NED are distributed in geographic coordinates at 1/3, 1, and 2 seconds of arc for any geographic location within the conterminous United States, Alaska, Hawaii, Puerto Rico, the U.S. territorial islands, Mexico, and Canada.

7.1.5 Light Detection and Ranging (LiDAR)

LiDAR is a remote sensing technology used, among other things, to examine the surface of the Earth. It uses laser technology to send signals from an aircraft to the target and analyze the reflected light. These light pulses—combined with other data recorded by the airborne system— generate precise, three-dimensional information about the shape of the Earth and its surface characteristics.[9]

LiDAR equipment principally consists of a laser, a scanner, and a specialized GPS receiver. Airplanes and helicopters are the most commonly used platforms for acquiring LIDAR data over broad areas. [3]

---

8. Origin of datum was changed from Meades Ranch in Kansas (NAD27) to the Earth's center of mass (NAD83), for compatibility with satellite systems.
9. While LiDAR systems were used by the federal government as early as the 1960s, primarily for atmospheric studies, it wasn't until 2000 that large public LIDAR data-gathering projects started.

7.2 Communications Equipment Installation

Telecommunications equipment comes in all shapes and sizes. One of the key requirements for ensuring reliability of the communications system is the environment in which the communications equipment is installed. To improve the uptime for equipment operation and enhance the life of equipment, it is important to carefully consider the requirements of the equipment enclosure.

For utility purposes, communications equipment can be installed in a control room (inside of the electric substation), an outside shelter for radio repeaters and fiber-optic regenerators, cabinets mounted on the electrical poles and towers, or enclosures buried in the ground. Each of them will have their own set of requirements, briefly described in the following sections.

7.2.1 Equipment Rooms, Shelters and Cabinets

To properly protect communications equipment, shelters are designed and often custom-built to house it. Different materials can be used to construct the shelters, including metal frames, concrete, wood frames, and lightweight insulated materials. Each type of material offers benefits for particular environmental conditions and site locations. Some of the design requirements are described as follows:

- Equipment: The shelter has to be properly designed for the electrical and mechanical requirements of the proposed equipment.
- Backup power: Backup power sources, such as uninterruptible power supply (UPS), DC power plant, and generators have to be correctly sized to handle the designed equipment loads.
- Climate control: It has to be sized to match the local climate and equipment generated heat loads, and provide optimal equipment environmental conditions [heating, ventilation and air-conditioning (HVAC)].
- Local and remote alarms: Alarming capabilities have to be adequately designed for monitoring, diagnosing, and alerting about problems such as failures of critical components and intrusion.
- Surge suppression: Grounding and transient voltage surge suppression systems have to be designed to provide protection from high-voltage anomalies.
- Installation: The equipment has to be installed in the most efficient configuration, allowing for access and maintenance, while meeting the manufacturer recommendations. For example, microwave radio racks should be placed as close to the waveguide entrance as possible in order to avoid unnecessary twists and bands of the waveguide(s).
- Cable management: Cable entry and routing facilities have to be properly designed for the proposed equipment interconnections.
- Optional and/or customized equipment shelter features: Such features include multistory building design, seismic/wind-rated design and reinforced building structure, lightweight fiberglass or metal construction for rooftops or

temporary installations, FM200 Fire suppression[10], indoor generator rooms, alternate fuels for the generator, off-grid power sources (solar and/or wind), waveguide entries, remote alarming capability, passive cooling, explosion-proof design (used commonly in oil refineries), bullet-resistant doors and/or bullet-resistant walls according to the ballistic UL 752 standard[11], and special visual appearance design according to local law and bylaw requirements.

Cabinets have been used for housing communications equipment for many years and have recently become the option of choice for copper, radio, and fiber-optic equipment. They are relatively small, with efficient use of rack space for mounting the payload as well as the power plant.

Large equipment rooms and control centers quite often have to comply with the *network equipment-building system* (NEBS) standard, which is the most common set of safety, spatial, and environmental design guidelines developed in the United States and applied to communications equipment. Today, NEBS requirements are utilized all over the world for a host of commercial, utility, and defense applications. Telcordia, and its predecessor, Bellcore, have maintained the NEBS documents since the divestiture of the Bell System in 1984.

NEBS level 1 addresses the personnel and equipment safety requirements of GR-63-CORE and GR-1089-CORE.

NEBS level 2 addresses equipment operability in a controlled environment (usually data centers) that will not be subjected to environmental stress. Levels 1 and 2 are aimed at relatively benign environments and are rarely used.

NEBS level 3 is a term from Bellcore special report, SR-3580, and means the equipment meets all of the requirements of GR-63-CORE and GR-1089-CORE. NEBS level 3 has strict specifications for harsh environments, like fire suppression, thermal margin testing, vibration resistance (earthquakes), airflow patterns, acoustic limits, failover and partial operational requirements (such as chassis fan failures), failure severity levels, RF emissions and tolerances, and testing/certification requirements.

Aside from having individual pieces of equipment be NEBS-compliant, the complete solution itself must be tested to determine if it complies with the standards. The connections and interactions between the pieces of equipment may result in a solution that fails to meet the standards. Any changes made in the equipment may require retesting of the equipment or the entire solution.

10. FM-200 is a clean, colorless, and environmentally friendly fire suppression agent that is electrically nonconductive and safe for humans. It extinguishes flames primarily through heat absorption, leaving no residue after a fire.

11. UL 752 defines bullet resistant products according to eight distinct levels of ballistic protection, each capable of withstanding a different size of ammunition and Level 4 is a commonly referenced standard for telecommunications shelters. There are a number of other standards for rating of bullet-resistant materials, like National Institute of Justice (NIJ) 018.01, State Department SD-STD-02.01, ASTM F-1233, European Standard DIN EN 1063, British Standards Institution BS 5051, German Deutche Institut fur Normung (DIN) 52-290, Councils of Standards Australia / New Zealand AS/NZ 2343, etc.

7.2.2 Equipment Racks

Telecommunications equipment comes in all shapes and sizes, so equipment shelters that house such equipment are often custom-designed and custom-built to properly protect it. Figure 7.1 represents typical communications racks in a substation control room.

Since space is always at a premium, one of the basic assumptions (that needs to be verified from existing drawings and/or by site visit) is that there is available space to install the microwave radio, antenna, and/or fiber-optic multiplexer and that all other miscellaneous equipment, and that AC power (and maybe even a DC charger and battery backup) is readily available.

On simple radio systems, the same team will install radio, coax cables, antennas, and even the power supply. On microwave systems, a radio supplier usually provides people to install radios, but large antennas, waveguides, pressurization equipment, and so forth are installed and tested by a different team of people who specialize in this kind of work.

Fiber-optic cables coming into the equipment room should be terminated at the fiber patch-panel and as close to the entrance point as possible. They are usually brought into the building in the PVC duct and separately protected by inner duct(s). Installers have to closely adhere to fiber-optic cable manufacturer recommendations regarding the maximum bending radius of the cable.

7.2.3 Global Positioning System (GPS) Antennas

The purpose of the GPS is to provide users with the ability to determine their location in three-dimensional space or for synchronization of IEDs.[12] In addition, many utility sites contain equipment that requires timing provided by the GPS. To accomplish these functions, the receiver must be able to lock onto signals from at least four different satellites. Moreover, the receiver must maintain a lock on each satellite's signal long enough to receive the information encoded in the transmission [4].

Achieving and maintaining a lock on four or more satellite signals can be impeded because the signal is transmitted at 1.575 GHz, a frequency that is too high to bend around or pass through solid objects in the signal's path. This is why GPS receivers cannot be used indoors. Outdoors, tall buildings, dense foliage, and terrain that stand between a GPS receiver and GPS satellite will block that satellite's signal.

An optimum site for GPS antennas provides unobstructed view of the sky in a hemisphere around the antenna base, with no obstruction above a 10-degree vertical angle, measured from the horizontal plane running through the antenna base

12. Currently, two Global Navigation Satellite Systems (GNSS) are operational: the American Global Positioning System (GPS) and the Russian Global Navigation Satellite System (GLONASS). These so-called Medium Earth Orbiters (MEOs) transmit signals in two frequency bands, L1 and L2, and will soon transmit in a third band, L5. The L1 frequency is 1,575.42 MHz; the L2 frequency is 1,227.60 MHz; and L5 is 1,176.45 MHz. Envisioned as an alternative and a complement to the U.S. Global Positioning System (GPS) and the Russian GLONASS, Galileo is a GNSS currently being built by the European Union (EU) and European Space Agency (ESA). The fully deployed Galileo system will consist of 30 satellites and the associated ground infrastructure, and it should be operational by 2020.

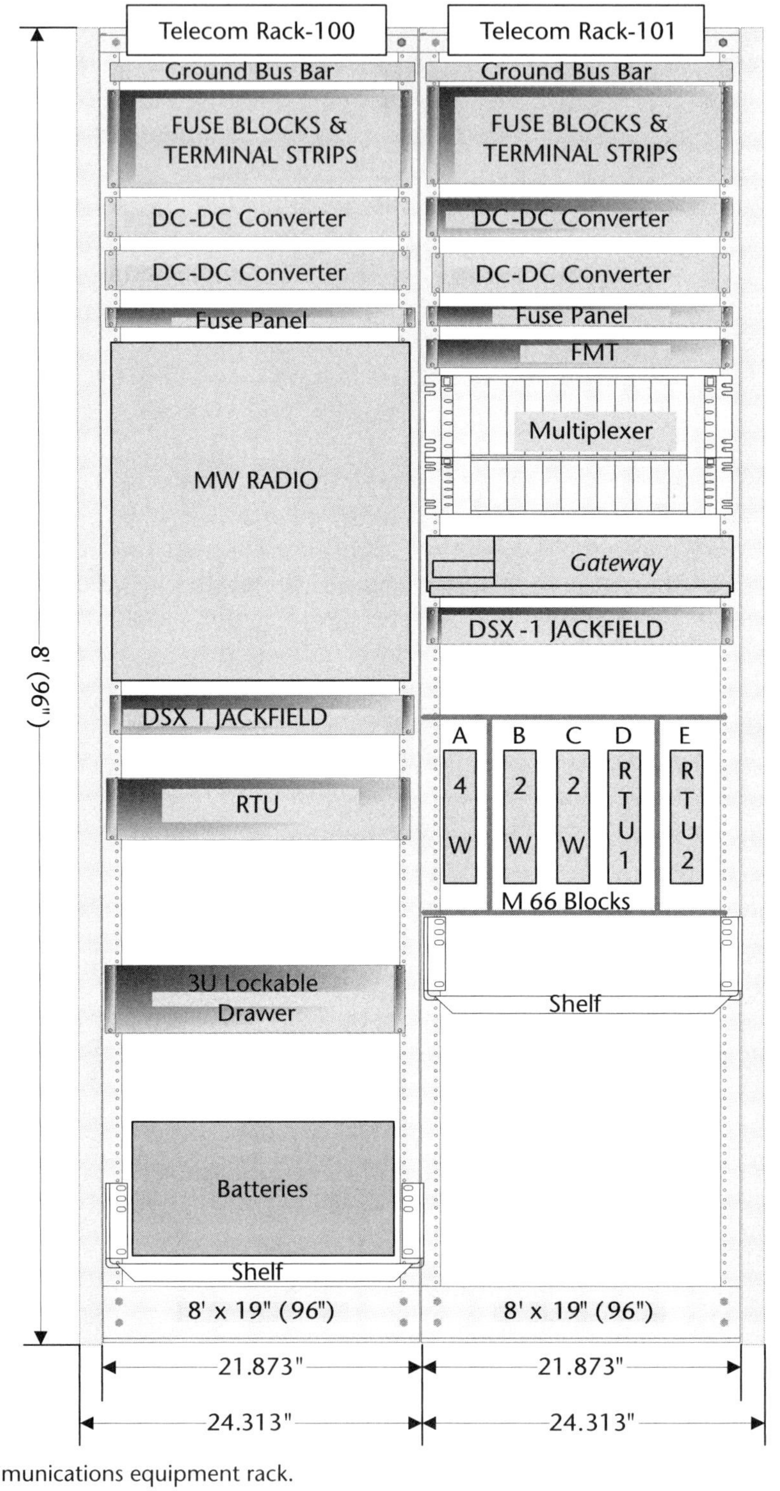

Figure 7.1 Communications equipment rack.

and a line of sight from the center of the antenna to the top of an obstruction.[13] Self-support towers or guyed towers will not generally block the GPS signal and high-voltage line will not cause interference with GPS signal reception.

7.3 Cables and Connectors

There are literally hundreds of different types of cables and connectors in use, depending on their application and environment in which they are installed. The required length, data rate, frequency, and reliability will all determine what kind of cables and connectors are going to be used for a specific application. For cable systems inside substations see [5].

7.3.1 Twisted Pair and Ethernet Cables

Years ago, copper wire was the only means of transporting information. Technically known as UTP, such wire consisted of a large number of pairs of copper wire of varying size in a cable. The cable did not have a shield, and therefore the high-frequency part of the signal was able to leak out. In addition, the twisting on the copper pair was very casual, designed mostly to identify which wires belonged to a pair as to handle transmission problems. Even with these limitations, it was quite satisfactory for use in voice communications.

As networks transitioned to twisted-pair Ethernet, unshielded twisted-pair cabling became the standard, especially in the United States. UTP cable has been and still is relatively inexpensive and easier to terminate and install (for the most part) than shielded cable. Shielded cable typically is also used only in noisy environments with a lot of EMI/RFI (and in high-security type venues, such as finance and government). Shielded cable is more common in European nations.

F/UTP denotes foiled/unshielded twisted pairs, which consists of four unshielded twisted pairs encased in an overall foil shield. Other names for this type of cable are FTP (foiled twisted pair), ScTP (screened twisted pair), or S/UTP (screened/unshielded twisted pair). There is also an S/FTP (screened/foiled twisted pair) cable, normally a CAT7 cable that has four individually shielded twisted pairs encased in an overall braided shield.

Category 3 (CAT3) cable is rated for networks operating up to 16 Mbps. It is suitable for voice transmissions (not VoIP). ISO/IEC refers to the end-to-end channel as Class C. *Category 4* (CAT4) cable is rated for transmissions of 16 Mbps up to 100m and today considered obsolete. *Category 5* (CAT5) cable was common for 100-Mbps LANs, ratified in 1991, and is today considered obsolete.

Enhanced Category 5 (CAT5e/Class D) cable, ratified in 1999, was designed to enable twisted-pair cabling to support full-duplex, 100-MHz applications such

13. The signals from the GPS satellites previously were degraded intentionally by the U.S. Department of Defense (DoD) for the purpose of national security. This performance degradation is known as selective availability (SA), and only DoD-approved users had access to satellite signals without SA. However, a policy statement issued by the White House indicates that SA was turned off in 2000, and the accuracy of GPS position fixes has been improved significantly. At this time there is no selective availability in force although it can be reactivated at any time without the notice to GPS users.

as 100BASE-TX and 1000BASE-T.[14] CAT5e introduces stricter performance parameters such as power-sum near-end crosstalk (PS-NEXT), equal-level far-end crosstalk (EL-FEXT), and power-sum equal-level far-end crosstalk (PS-ELFEXT).

Category 6 (CAT6/Class E) cables easily handles gigabit Ethernet (1000BASE-T) applications. It's a 100-ohm cable with a frequency of 250 MHz. *Augmented Category* 6 (CAT6a/Class EA), a relatively new standard, is designed to meet or exceed the requirements of 10-gigabit Ethernet over copper at 100m. It extends the frequency range of CAT6 from 250 to 500 MHz. Category 6A is also designed to support bundled cable installations up to 100m and PoE+ low-power implementations and introduces new *alien crosstalk* (ANEXT) measurement requirements.

Category 7/Class F is only an ISO/IEC 11801:2002 standard and is not in a draft stage by the TIA. It's designed to meet or exceed the requirements of 10-gigabit Ethernet. The standard specifies a frequency of 1–600 MHz over 100m of fully shielded twisted-pair cabling.

7.3.2 T1 Cables

Copper cable used to transmit T1 signals has been available for many years. A typical design has been two-pair 22AWG with foam skin insulation, each pair shielded, with 100-ohm impedance (commonly known as ABAM telephone cables).

The physical size of these cables requires termination with DB-9 connectors, DB-15 connectors, punch-down hardware or mechanical terminal blocks. These older cable designs and their termination systems were developed for T1 applications within a telephone company's CO facility. Now that T1 circuits are finding their way into LAN wiring closets and other confined spaces, the requirements have changed. Most new DS1 system interface devices utilize either shielded or unshielded RJ 45 telco modular connectors. The older ABAM cables are too large to fit these relatively inexpensive and widely used connectors.

The local telco service supplier usually provides T1 service into the entrance facility or to the "demarcation point," and then the installer must take it from there. Since many LAN installers and owners are unfamiliar with the requirements for T1 cabling, many use the wrong cabling; for short runs (less than 30 or 40 feet), just about any 100-ohm shielded cable will work, but longer runs of incorrect cables may cause equipment to fail.[15]

7.3.3 Coaxial Cables

Today, coaxial cable technologies are primarily used for connecting antennas to the radio transmitter as well as a connection between an indoor (IDU) and outdoor (ODU) radio unit on a split-configuration microwave radio system. Coax cables are much easier to install, and they are presently used more often than waveguides.

---

14. The pair twists are responsible for much of cable's performance. When terminating CAT5e or higher, maintain pair twists to within 0.5 inch (13 mm) from the point of termination and remove as little of the sheath as possible.

15. Many installers do not realize that the cabling must support specific DS1 system requirements that are defined by an ANSI specification. These include BERs, EMI shielding, a pulse shape definition and many others. In short, DS1 systems are very different from LAN Ethernet systems.

For coaxial cables interconnecting IDUs and ODUs [operating at intermediate frequency (IF) of a few hundred megahertz], the maximum acceptable length is vital information for the installation. Usually, the maximum length is about 300m (984 ft) due to the electrical limitations of the coaxial cable itself.

The IDU-ODU coaxial cable should be grounded, with cable grounding kits, to the tower or station Earth to minimize the potential for damage due to lightning or other externally induced interference. It should be grounded as a minimum at the base of the tower and close to the transition into the equipment shelter. Grounding this cable close to the ODU will provide further protection, although it is not essential in areas of low lightning risk. Providing lightning protection is essential to achieve proper equipment protection.

Most antenna problems are caused by coaxial cable connections that loosen due to vibration, allowing moisture to penetrate the connector interface, so it is recommended that all outdoor cable connections be weatherproofed.

7.3.4 Waveguides

Elliptical waveguides are hollow copper tubes, and their installation is a very tricky job that must be performed by trained and experienced technicians. Very small deformations can introduce enough impedance mismatches to produce serious problems. Waveguides are typically offered in three versions: standard, premium (low-VSWR) and overmoded (low-attenuation) and operate typically from 2 to 11 GHz. Standard waveguides are recommended for low- and medium-capacity microwave systems, while premium elliptical waveguides are recommended for high-capacity microwave systems.

Elliptical waveguides may be ordered with the connectors attached and tuned if the precise length is known in advance. If the connectors are to be installed in the field, it may be necessary to sweep and tune the waveguide and connectors to realize optimal return loss. The waveguide must be supported with special hangers at regular intervals to prevent stress, movement, and excessive pulling forces caused by the cumulative weight of the waveguide. Hangers are usually installed at 3-ft (1-m) intervals.

A waveguide must be grounded using special grounding kits. It is common practice to ground the top of the waveguide run, the bottom of the run, and the radio terminal. For waveguides making the connection from the indoor branching to the outdoor antenna, the important specification is the minimum bending radius (E-plane and H-plane).

The waveguide (and air dielectric coaxial cables) must be pressurized with equipment designed to dry the pressurized air. This action is necessary to displace any water vapor from the waveguide or cable that will significantly increase attenuation. Since this is a closed system, pressurization equipment operates (turned on) very rarely; increased activity of the compressor could be an indication of the leak in the system and requires troubleshooting.[16]

16. Unfortunately, parabolic antennas and transmission lines in remote areas are quite often a target for practice shooting, and holes in the transmission lines or damaged antenna feeder systems are a common result of such a barbaric act.

7.3.5 Impedance Matching and Return Loss Measurements

Impedance, in antenna terms, refers to the ratio of the voltage to current (both are present on an antenna) at any particular point of the antenna. For an efficient transfer of energy, the impedance of the radio, the antenna, and the transmission line (cable or waveguide) connecting them must be the same. Transceivers and their transmission lines are typically designed for 50-ohm impedance. If the antenna has impedance different from 50Ω, then there is a mismatch and an impedance matching circuit is required. Of course, other impedances are used as well.

The reflection coefficient, Γ, is calculated as:

$$\Gamma = \frac{Z_L - Z_0}{Z_L + Z_0} = \frac{\frac{Z_L}{Z_0} - 1}{\frac{Z_L}{Z_0} + 1} \tag{7.1}$$

where Z_0 is a characteristic impedance of the transmission line, and Z_L is a load impedance. Expressed in logarithmic forms, the reflection coefficient is called *return loss*:

$$R_L = -20 \log \Gamma \tag{7.2}$$

Return loss is another way of expressing impedance mismatch. It is a logarithmic ratio measured in decibels that compares the power reflected by the antenna to the power that is fed into the antenna from the transmission line. For example, if the antenna absorbs one half of the signal, and one half is reflected back, we say that the return loss is –3 dB. An antenna might have a value of –10 dB (90 percent absorbed and 10 percent reflected).

The *voltage standing wave ratio* (VSWR) [also known as the standing wave ratio (SWR)] is defined as the ratio of the magnitude of the maximum voltage on the line to the magnitude of the minimum voltage on the line, and it can be expressed as:

$$VSWR = \frac{1 + \Gamma}{1 - \Gamma} \tag{7.3}$$

VSWR is a measure of impedance mismatch between the transmission line and its load; the higher the VSWR, the greater the mismatch. The minimum VSWR (i.e., that which corresponds to a perfect impedance match) is unity. The combination of the original wave traveling down the coaxial cable (toward the antenna, or the opposite during receive) and the reflecting wave is called a *standing wave*.[17] The ratio of these two waves is known as the SWR.

Because the SWR is not always calculated from the voltages, the "V" is sometimes dropped from the acronym and it is referred to as the standing wave ratio (SWR). VSWR and SWR are the same number and can be used interchangeably.

17. In other words, superposition of incident and reflected wave forms a standing wave. A standing electromagnetic wave does not propagate, but instead the electric and magnetic fields execute simple harmonic motion perpendicular to the would-be direction of propagation.

Table 7.1 Impedance Mismatch

Z_L/Z_0	*Reflection Coefficient,* Γ	*Return Loss [dB]*	*VSWR*	*Comment*
Infinity	1.00	0.00	Infinity	Open Circuit
100.00	0.98	0.17	100.00	
10.00	0.82	1.74	10.00	
5.85	0.71	3.00	5.85	Half Power
3.00	0.50	6.02	3.00	
2.00	0.33	9.54	2.00	Good Match
1.20	0.09	20.83	1.20	Excellent Match
1.00	0.00	Infinity	1.00	Perfect Match
0.83	–0.09	20.83	1.20	Excellent Match
0.50	–0.33	9.54	2.00	Good Match
0.33	–0.50	6.02	3.00	
0.17	–0.71	3.00	5.85	Half Power
0.10	–0.82	1.74	10.00	
0.01	–0.98	0.17	100.00	
0.00	–1.00	0.00	Infinity	Short Circuit

A VSWR[18] of less than two is considered a good match. (See Table 7.1.) However, we assume here that the transmission line and the load connected to that line have purely resistive impedance, in other words characteristic impedance of the transmission line, Z_0, and load impedance, Z_L, are real numbers. This is a simplistic approach, but sufficient for this discussion and most practical applications.

The characteristic impedance of the industrial low-loss transmission lines has a constant value, normally 50 or 75 ohms, over a large frequency band, even when the line loss is not zero.

Impedance matching is a process of making the output impedance of a source equal to the input impedance of the load in order to maximize the power transfer and minimize reflections from the load. Mathematically, it means that the load impedance is the complex conjugate of the source impedance. The impedance for the entire chain from the radio to the antenna must be the same, and almost all radio equipment is built for an impedance of 50Ω.

VSWR/return loss can be measured for the antenna, transmission line, transmission line and all rigid components, and the entire antenna subsystem (e.g., antenna, transmission line, and connectors). These measurements ensure correct impedance matching between the radio port and antenna port. In addition, the insertion loss of the transmission line and all other components should be measured.

Return loss measurement of just the transmission line may be performed by placing an ideal termination at the opposite end of the transmission line. Insertion loss may be measured by shorting the transmission line. It is necessary to use a shorting connector for coaxial cable and a shorting plate for the elliptical waveguide.

18. One common way of visualizing the VSWR is to use a polar plot called a Smith chart. From this plot for the VSWR value, the return loss and the impedance for the different frequencies can be derived. Therefore, it is an important instrument for understanding antennas and transmission lines.

7.4 Antenna-Mounting Structures

7.4.1 Monopole, Self-Support, and Guyed Towers

There are many different options when it comes to mounting antennas and radio equipment. Small antennas can be mounted on small pipes attached to the equipment room, wooden poles in the ground, and tripods (as a temporary solution).[19] Large microwave antennas, like parabolic dishes, need a large communications tower made of steel—a monopole, self-support, or guyed tower. In addition, electric transmission towers and distribution poles are sometimes used as antenna-mounting structures.

Structure type and size will depend on the size of planned antennas, the number of outdoor radio units, and the waveguides and cables to be installed, as well as the antenna height required for achieving clear LoS.

Ranging in height from 25 to 125 ft (7.6–38m), *monopoles* consist of a single pole, approximately 3 ft (1m) in diameter at the base, narrowing to roughly 1.5 ft (0.5m) at the top, and they may support any combination of whip, panel, or dish antennas. The efficiency of the design allows it to be an economical option for a range of heights of up to 40m (130 ft).

Self-supporting towers are most commonly used towers in suburban and rural areas and require a base spread of approximately 13 percent of their height. The tower can be made using a variety of steel shapes and connection types. A common makeup is angle leg, angle bracing, and bolted construction, with a climbing ladder inside the tower.

The design, supply, and installation cost of a self-support tower is approximately twice the cost of an equivalent (between 15 and 76m in height) guyed tower; beyond 76m, the difference in cost is even greater.

Guyed towers are slender, generally three- or four-sided lattice in construction, and uniform in dimension over their length. They are maintained in their vertical position by guy wires, which are attached at various levels on the tower and anchored to the concrete on the ground. Under normal circumstances, a guyed tower requires a radius of 80 percent of the height of the tower and therefore needs a considerably larger ground area than a self-support tower, which might be costly.[20] Guyed towers often can be built very quickly and are much lighter because less construction material is required. Hence, they have a smaller wind load area.

7.4.2 Structural and Other Requirements

The construction standard for today's communications towers is primarily modeled after American National Standards Institute (ANSI) standards. Communications towers are designed in accordance with "Structural Standards for Steel Antenna Towers and Antenna Support Structures," ANSI/TIA/EIA-222-G (released in

19. Minimum visual impact antenna structures, for example, a manmade tree, clock tower, church steeple, bell tower, utility pole, light standard, identification pylon, flagpole, or a similar structure are often used by wireless companies but rarely by power utilities. They are designed to support and camouflage or conceal the presence of telecommunications antennas.

20. Guyed towers are completely prohibited in some countries (Germany, for example), demanding self-support towers or monopoles.

2006).[21] The ANSI standard is based on equations developed by professional engineers using wind tunnel testing to accurately predict the effect that wind has on telecommunications structures.

Most structures are designed to withstand a forceful wind speed that occurs on the average of once every 50 years. This wind speed is then escalated, with height, to a much higher wind speed at the top of the structure. A gust factor to account for the varying nature of wind is also incorporated into the design of the structure. The ANSI standard lists minimum 50-year return wind speeds for all counties within the United States.

Radial ice accumulation is also accounted for in the design. Ice loads are escalated with height since ice accumulation is known to increase with wind speed.

In Canada, tower strength and design are not regulated, so Canadian Standard S37, "Antenna-Supporting Structures—A National Standard of Canada," is not a legal requirement.

Some important topics to consider in civil construction are quality control, construction of foundations, concrete reinforcement methods and anchor design, and quality of paint and welding points. The tower/mast is painted mainly for corrosion protection. A compulsory aviation warning coloring scheme is deployed in some countries while aesthetic requirements must be considered in other countries.

Different types of aviation warning lights are required in various countries, including dusk-activated relays. It is widely accepted that lightning protection can be connected to Earth through the tower/mast construction.

A maximum resistance of 10Ω to ground is dictated by British standards and is also widely observed elsewhere. However, in some countries (including the United States), more stringent criteria are sometimes applied.

7.4.3 Maximum Allowed Antenna Deflection

Generally speaking, most structures are limited to a 4-degree twist or sway rotation and a horizontal displacement equal to 5 percent of the height of the structure. More stringent rotation requirements are usually provided for structures supporting microwave antennas. Figure 7.2 illustrates problems due to twist and sway of the antenna structure.

Although in most cases microwave engineers like to install microwave antennas as high as possible from the LoS perspective, it may be more desirable to mount them as low as possible from the twist and sway perspective.

For microwave links, a typical limitation in twist/sway for the structure (tower and antenna) corresponds to a maximum 10-dB signal attenuation due to antenna misalignment. A simple formula for estimating the maximum allowed deflection (single side) in degrees, for microwave transmission antennas is

21. A nationally recognized standard jointly sponsored by the Electronic Industries Alliance (EIA, formerly the Electronic Industries Association) and the Telecommunications Industry Association (TIA). While it is not mandatory to upgrade existing towers when new editions of the Standard are published, communication structures are frequently subject to changes in attached equipment. This necessitates conformance with the current edition.

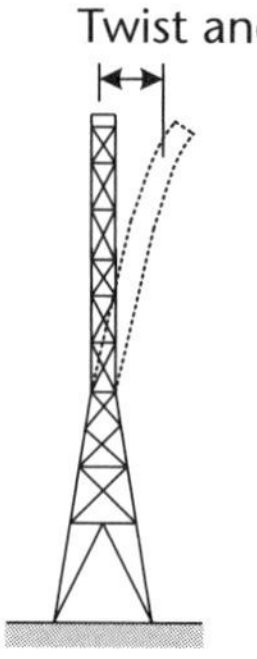

Figure 7.2 Twist and sway of the tower.

$$\alpha_{-10\text{dB}} = \frac{18}{f \cdot D} \ [°] \tag{7.4}$$

where

D = antenna diameter (m);

f = frequency (GHz).

So, for example, maximum allowed 10-foot (3 meters) dish antenna deflection at 6 GHz would be approximately 1 degree.

7.4.4 Tower Procurement, Installation, and Maintenance

It is the designer's task to make each tower as small as possible given the antenna and transmission line loading, LoS, code, deflection, and other requirements. A reputable tower design/manufacturing company is a good source of reliable information regarding pricing, scheduling, and engineering. Consulting firms that specialize in towers are another source of unbiased, objective information on creating a specification for towers and tower design loading.

The most vital information that has to be provided to the tower manufacturer is described as follows:

- Tower load: Antenna-mounting structures include antennas (immediate and future requirements), wind, ice, waveguides, coax cables, platforms, the waveguide bridge, and microwave outdoor radio units (ODUs).
- Wind speed: Also referred to as wind load, this is the force the wind has on the tower and antennas (survival and operation).
- Ice load: Also known as radial ice, this is the amount of ice in inches formed around each tower member.
- Soil report: Details the soil conditions present at the site and helps to determine what type of foundation is required.
- Other design specifications: Include FCC, FAA, and others. For example, in the United States, the paint color and lighting requirements on towers over 200 feet, or within five miles of an airfield, are strictly controlled by

the FAA[22]. In Canada, painting of the towers must comply with Transport Canada Standard 621 [6].

Many towers are protected against corrosion by a hot-dip galvanizing process. This process is regulated in accordance with ASTM A123/A153 and CSA Standard G164 to ensure quality standards. Galvanization is a long lasting (25 years in coastal locations and 50 years in urban locations), typically only requiring minor repairs due to scratches accumulated during tower delivery and stacking, or from installation of additional hardware such as antenna and cellular sector mounts.

The TIA-222-G provides communications tower design guidelines but it also outlines a recommended schedule for tower inspections, typically every three years for a guyed tower and five years for a self-support tower. Guyed towers require more frequent maintenance than self-supporting towers, and the guy cables should be inspected for proper tension and to detect corrosion.

Most deformations (but not all) in the tower, which could be a sign of structural damage, can be seen from the ground; a visual structural inspection should include careful investigation of tower members and guy point foundations, guy point termination plates (including below grade), turnbuckles and sleeves, as well as the alignment of the guy wires and the tower itself. Structures located in hurricane-prone areas should be inspected annually for loose hardware and to ensure that antenna and transmission line attachments are secure.

Towers in coastal areas should also be inspected annually because they are subject to corrosive salt air and high winds. A good paint job will last five to eight years at the most and the maintenance of tower paint is crucial to the life of the structure.

The final part of any tower inspection should include a physical evaluation of the structure's various transmission lines, jumpers, connectors and antennas. Transmission lines (especially waveguides) should be inspected visually along their entire length for signs of damage or fatigue to the outer plastic jacket.

Within 30 minutes of the failure of any light on the tower or of either of the two incandescent bulbs in a beacon (slow flashing red light), a *Notice to Airmen*[23] must be filed with the FAA. The FAA is more lenient for steady burning marker lights, only requiring replacement when more than one marker is out at any marker level.

7.5 Power Supply and Battery Backup

7.5.1 AC Power

Although most telecommunications equipment requires the usual telco power (–48 VDC), there is a chance that some additional external equipment may require AC power. There are three distinct AC uninterruptible power supply (UPS) configurations: standby, line-interactive, and online.

22. FAA, Circular AC 70/7460-1K, 2007.

23. The International Notices to Airmen feature significant international information and data which may affect a pilot's decision to enter or use areas of foreign or international airspace. The entire document is published on the internet at the following address: http://www.faa.gov/air_traffic/publications/notices/.

A *standby UPS* (also known as an *off-line UPS*) consists of a basic battery power-conversion circuit and a switch that senses irregularities in the electric utility. The load is connected directly to the utility power, and power protection is available only when there is an outage. However, some standby UPS units include suppression circuits or power line conditioners to increase the level of protection they offer.

A *line-interactive UPS* offers a higher level of performance by adding voltage-regulation features to conventional standby designs. Like standby models, line-interactive UPS units protect against power surges by passing the surge voltage to the load until it hits a predetermined limit. At this point, the unit switches to the battery, but only after passing most surges through to the load. Line-interactive units can provide moderate protection against high-voltage spikes and high-frequency transients, but they do not provide complete isolation for the load.

An online AC UPS uses double conversion (AC/DC and DC/AC), providing complete isolation from most types of power problems.

7.5.2 DC Power

Communications equipment is usually required to operate off the –48- and/or +24-V power supply. Although most telecommunications equipment is located in places like control rooms and communications rooms, unattended telecommunications equipment in cabinets, huts, substations and controlled environment vaults (CEVs) are often subjected to harsh electrical and environmental conditions. Equipment, particularly batteries, deteriorates very fast under those conditions, so remote monitoring systems are often recommended.

The rectifier has to be able to recharge empty batteries in *x* hours (usually 4 hours) while carrying the full load of the system indefinitely. Ideally, the system should be easy to upgrade and expand (modular approach) to accommodate changes and growth.

Power utilities use DC-DC converters quite often, and they can be configured as primary or back-up/redundant power sources. DC/DC converters operate from 12-, 24-, 48-, 60-, 125-, and 130-VDC power sources and they can also provide 12-, 24-, 48-, 60-, 125-, and 130-V isolated, regulated DC outputs ideal for powering telecommunications equipment, SCADA networks, and other electronics (Figure 7.3). They are typically designed for use with 19- and 23-inch equipment racks.

Sometimes, in order to increase reliability of the power plant, redundant system design (A + B) is used, where two independent rectifiers and battery systems provide half the repeater load in normal conditions. Should one system fail, the remaining system can power the entire site's load.

7.5.3 Batteries

The battery is by far the most unreliable part of any system that depends on it. This is true because, if any one of many external variables is incorrect, the natural deterioration process within each cell is accelerated. When the utility power fails and the battery instantly assumes the critical load, the choices and investments that have been made in emergency power equipment will be realized. The accelerating factors include excessive cell temperature and excessive many discharge cycles.

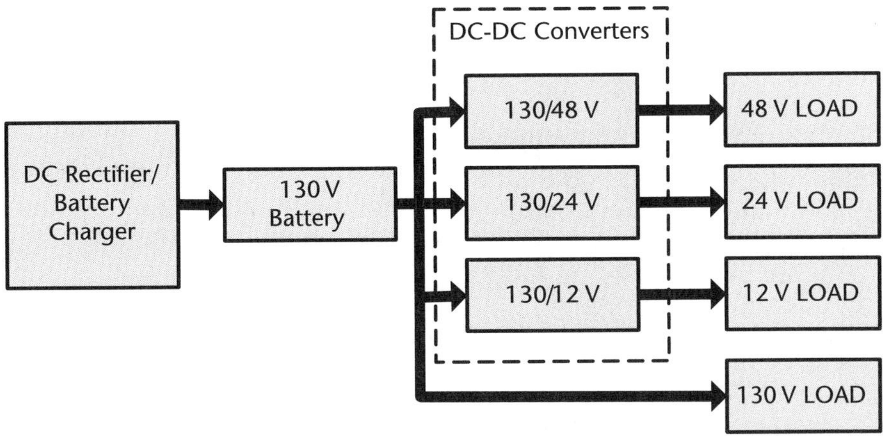

Figure 7.3 DC rectifier and DC-DC converters.

Historically, backup power for telecommunications systems has been lead-acid batteries; they come in two primary types, *flooded* and *absorbent glass mat* (AGM) batteries.[24] Flooded batteries are generally too large to fit into the limited space available for communications sites applications. Service and maintenance of flooded batteries in remote sites would also be very expensive, since the service technicians would have to visit many sites on a regular basis to check electrolyte levels.

Over the years there has been a trend toward other types of batteries, such as *valve-regulated lead-acid* (VRLA)[25] batteries (also called *gel cell* batteries). The term *valve-regulated* identifies a battery that is equipped with mechanical safety vents that can open under excessive overcharge.

Equipping power plants with a remote monitoring and control system is a very smart investment, as it reduces both the cost and frequency of personnel visits to the sites. They usually generate alarm messages but also can provide vital data about power plant conditions.

7.5.4 Motor Generators and Fuel Cells

Battery backup is an efficient solution for telecommunications sites where short power interruptions are expected, from a few minutes up to a few hours. In cases where grid power is unreliable or even nonexistent (remote radio sites and/or microwave repeater sites), alternative solutions to power the site have to be implemented.

Propane, diesel, or natural gas motors that turn electric generators provide higher-power solutions. Motor generators can be used in hybrid power systems, in combination with solar, battery, and/or wind. They require periodic maintenance (as well as regular fuel delivery) and have higher operating and lifetime costs.

Fuel cell technology is a new option in extended runtime backup power to replace or enhance battery or diesel generator systems. Fuel cells are electrochemical devices that convert the chemical energy of a reaction directly into electrical energy.

24. Absorbent glass mat (AGM) is a newer class of battery in which the electrolyte is absorbed into a mat of fine glass fibers. AGMs are more expensive than flooded (liquid) batteries but offer very good reliability.

25. VRLA battery was invented in 1859 by French physicist Gaston Planté, and represents the oldest type of rechargeable battery. VRLA batteries cannot spill their electrolyte.

Lower maintenance, lower fuel costs and emissions, as well as longer life, are some of their advantages when compared to traditional diesel generators.

There are several different types of fuel cells, each using a different chemistry. A hydrogen fuel cell uses hydrogen[26] as its fuel and oxygen (usually from air) as its oxidant. Other fuels include hydrocarbons and alcohols or chlorine and chlorine dioxide.

To produce a usable amount of electricity, multiple fuel cells, each producing up to about one volt, are combined into a fuel cell stack. The stack is designed to produce the desired electrical current and voltage for a specific application. The direct current electricity produced can be varied over a wide range by altering the area and number of cells in the stack [7].

7.5.5 Off-Grid Energy Sources

Solar energy is available almost everywhere in the world and suitable for power supplies everywhere except maybe the polar regions. Solar panels do not require fuel supplies, have no moving parts, and require very low maintenance. Battery plant at remote sites should be designed for 10–30 days of battery autonomy, which is recommended for most remote solar-powered systems (low Arctic locations use 60–90 days autonomy).

Many remote sites use photovoltaic cells to generate power for onshore and offshore traffic control systems, microwave radio repeaters, and so on. *Photovoltaic* (PV) power is a semiconductor-based technology that involves converting light energy directly into an electric current that can either be used immediately or stored (e.g., in a battery) for later use.[27]

In remote sites solar energy can be used to replace the electrical utility–based power supply entirely (where conventional electricity is unavailable or impractical, or just as an additional power that will help during the peak hours, reduce electrical power consumption, and provide a backup in the event of grid failure). In either case, for every project, a very extensive study must be done to determine the feasibility and cost efficiency of such a system.

In *wind turbines* wind turns the turbine, generating electricity from the electrical generator, and DC output charges the battery. They are typically low-maintenance, requiring only occasional checks for damage to turbine blades, controller and regulator tests, charging tests, and battery tests.

One of the most crucial pieces of information needed when evaluating the wind energy potential of any given area or site is a reliable definition of the wind resource. A site with an average wind speed of lower than 9 mph would not benefit year around from a wind-powered installation. Some of the smallest wind turbines can be mounted on existing telecommunications towers and produce about 8–10 kWh at wind speeds around 9 mph.

26. It is important to keep in mind codes and safety standards for hydrogen storage and siting. Refer to National Fire Protection Agency (NFPA), Hydrogen Technologies Code, 2011, especially if the site is in proximity of other manmade structures.

27. Photovoltaic solar cells produce electricity directly from sunlight, and they are usually made of silicon. Through the photovoltaic effect, a typical 4-inch silicon solar cell produces about 1W of DC power.

Small wind and solar panels are often seen as complementary technologies, as both are used together as a hybrid system to offset the variable resources of the sun and wind. Generally, when it is sunny it is calm and when it is cloudy it is windy. This complementary effect is even greater during seasonal changes; during the winter and spring when the sun is at its least exposure, the wind is at its highest. It is possible to find the solar *insolation*[28] or average wind speeds for any area of interest on Earth by going to the NASA web site. [8]

7.6 Practical Aspects of Project Management

7.6.1 Definition of Project Management

Project management is the application of knowledge, skills, tools, and techniques to project activities so as to meet (or exceed) customer needs and expectations from a project. A project is defined as an undertaking of a nonroutine (unique), nonrepetitive nature having prescribed objectives in terms of scope, time, quality, and cost. Meeting or exceeding customer needs and expectations invariably involves balancing competing demands, scope, time, cost, and quality [9].

The various phases of a communications project may be defined differently by different organizations, but they generally fall into the following categories:

- Concept (or feasibility) analysis;
- Network planning and preliminary design;
- Detailed design and engineering;
- Deployment (also called implementation);
- Testing and commissioning.

While project management skills are quite distinct from engineering design skills, the requirements of good project management are not different from the requirements of good engineering or good management. Project management establishes nonconflicting expectations by coordinating the client's desires with the contractor's abilities—and when the unexpected happens, project managers adjust schedules and expectations.

7.6.2 Project Management Activities

Project management services are applicable in all phases of a project, from the initial concept through implementation, to the final commissioning and handover of an operational project [10]. It is therefore important that a client (which could be another department within the same organization or an outside entity) be aware of the full scope of services that can be provided. Project management services normally include certain basic activities such as the following:

- Planning and scheduling (i.e., time management);

28. Insolation is a measure of solar radiation energy received on a given surface area in a given time and comes from a combination of the words "incident solar radiation."

- Budgeting and estimating;
- Cost control and accounting;
- Quality control;
- Regular reporting.

In addition to the basic services, depending on the particular project, project management services may also include a number of other items such as structuring the project into easily manageable parts; preparing project policies and procedures; advising and assisting the client with respect to regulatory and permitting processes; dealing with consultants, vendors, and contractors; and managing construction/implementation/deployment for conformity with the approved design [11].

At the beginning of the project, the project manager will organize a *kick-off meeting* and invite all the parties involved in the project. This is a good opportunity for people to meet and see who else will be working on the project. In addition, this is a good time to identify any "missing links" in the project (i.e., equipment, experts, or anything else that might have been omitted by the project manager or previously thought not to be required). This is a brainstorming session organized by the project manager, but it requires active involvement by all team members.

Many projects, including communications systems, are a multidisciplinary activities that involve a number of specialized experts in their respective fields. The project manager will have to understand the basics of the communications system build-out process, know which specialists need to be brought into the project, and understand basic time-line and critical-path activities.

The *project system engineer* (also called the *lead engineer*) should be present at this meeting, since this person will be responsible for ensuring that the design and installation is performed according to industry, customer, and supplier standards and practices. The project system engineer is the key technical person on the project and has an overall responsibility for the technical integrity of the system design provided.

7.6.3 Communications Network Planning and Design

Network planning (also called preliminary design) is usually done at the beginning of the project and results of the preliminary network planning activities and its deliverables (budget, work force, and schedule) form the basis on which the whole project will be assessed and potentially approved or rejected. Of course, due to the short time available as well as very limited input data, the margin of error in these calculations and predictions is quite high. A few different scenarios (options) may be analyzed during the preliminary design (and/or feasibility study).

Detailed design consists of activities such as conducting site visits, performing link engineering for radio systems, optical budget calculations for fiber-optic systems, and defining need and locations for microwave repeaters and fiber-optic regenerators. During detailed design, regulatory activities (e.g., environmental, FAA, and FCC) have to be completed, as does the engineering of components such as shelters, cabinets, rack layouts, and cabling routes. The result is the final equipment specification [bill of materials (BoM)].

Results of the engineering process are forwarded to the companies that are providing communications equipment, towers, shelters, and other infrastructure, and this information will serve as input data for their scheduling, design, and production.

7.6.4 Equipment and Services Procurement

Procurement is the systematic purchasing of all materials, equipment, and services needed for the project, in good time, and in a manner that is cost-effective. These generally include (but may not be limited to) those provided by suppliers, consultants, installation and testing services, construction managers, and project management. Services may also include post-installation operation and maintenance of the communications network.

A decision is usually made either based on the *request for quotation* (RFQ) response and its compliance with the functional and technical requirements of the end user (owner). [12] The RFQ should be structured to simplify the process for the client and supplier alike; responses must specifically address each point set forth in the RFQ and should be clearly answered in the response tables. The amount of information submitted is left to the discretion of the respondent, but it is imperative that pertinent information be submitted and that individual topics of interest be dealt with completely and concisely. The RFQ response time varies, depending on the complexity of the project and client's requirements.

The proposals and/or quotes are usually based on uninterrupted/contiguous site field activities. Additional mobilization costs are usually not included in the pricing, and if the installation and other field work are delayed due to inclement weather (see *National Weather Service* for an accurate weather forecast in the area of interest) [13] an inaccessible site, or incomplete site preparation or construction by others, additional charges may apply.

Equipment evaluation is usually based not only on the technical specs and price, but also on the other criteria (e.g., the experience of other customers in the same or some other countries with the same or similar type of equipment, warranty, and customer support).

Cost-benefit analysis (CBA) is an important and widely used technique in decision-making. In many complex decisions, costs are set aside until the benefits of the alternatives are evaluated; discussing costs together with benefits can sometimes bring forth many political and emotional responses and therefore complicate the decision-making process to the point of no resolution.

To be more objective in this decision-making process, commonly some kind of decision table is created, and each vendor is evaluated based on a list of criteria. Criteria are weighted based on the importance to the end user, and the resulting spreadsheet will show the final ranking of the suppliers. Although this is a very common way of selecting suppliers, the first problem is that people are typically not very good at making complex decisions involving multiple criteria and multiple possible outcomes (alternatives). The second problem is that this final supplier ranking can easily be manipulated to fit the decision-maker's personal, biased choice.

A more objective decision-making process will most likely involve some type of analytical methods. *Analytical decisions* use the *multicriterial decision analysis*

(MCDA) methods to define priorities between alternatives (actions, scenarios, projects) according to multiple criteria, allowing for more in-depth analysis of the problem. MCDA methods rely partly on mathematical hypotheses and partly on information gathered from decision-makers and stakeholders. Different MCDA methods require different types of value information and could follow various optimization algorithms [14]. Some of the methods require a large number of participants and thus may not be suitable for the tendering process.

The analytic hierarchy process (AHP)[29] is a decision-making aid initially described in [15]. AHP[30] is suitable for complex decisions that involve the comparison of decision elements that may be subjective and therefore difficult to quantify.

Group decision is a way to combine different individual preferences on a given set of alternatives into a single collective preference. The expert auditors participating in making a group decision face the same common problem and are all interested in finding a solution or making a decision—no mathematical models can help with the resolution of completely opposed and/or possibly biased opinions. [16] Feedback and consensus among the participants in the decision-making process could be achieved, for example, by a *Delphi technique*[31] in which individual judgments are summarized and sent back anonymously to the rest of the group.

7.6.5 As-Built Documentation

An as-built document is usually not completed until each site is integrated; however, it should be stressed that an agreement of its content must be included in the contract at the beginning of the project. It is advisable to prepare a standard installation drawing or a real site and present it to the customer so as to minimize potential misunderstandings later.

The as-built document should include at least the following items: engineering documents; a site situation plan (showing the site locations on a map); floor plan drawings indicating the location of the installed equipment; cableway drawings, antenna placement information; an alarm allocation table; power distribution information; configuration data; the rack layout; plant specifications that list equipment used on the site, including installation material (e.g., cables); and a product list (including equipment serial numbers).

In addition, it should contain licenses to operate radio equipment (from, e.g., the FCC, the FAA, and the city), acceptance test documents, factory test results (description of the factory tests and test results performed at the manufacturer's site during the manufacturing and quality assurance process), and any other document that was specified at the beginning of the project.

29. It is easier and more accurate to express one's opinion on only two alternatives than simultaneously on many alternatives. AHP, developed by Saaty in the 1970s, uses pairwise comparisons to derive accurate ratio scale priorities, as opposed to using traditional approaches of assigning weights.

30. Application of AHP requires solid understanding of matrix algebra, so the methods and details of calculation process will not be discussed in details here since they are beyond the scope of this book. Nevertheless, it is useful to realize that there are very practical and applicable mathematical methods out there available to assist with a complex decision-making process.

31. Delphi technique was developed by the research group at the RAND Corporation, Santa Monica, CA, in 1969 to obtain the most reliable consensus of opinion from a group of individuals about a dilemma that had no objective resolution.

7.6.6 Maintenance Program

The maintenance program has to be established during the communications network planning phase. For example, for tasks such as replacing failed communications equipment or its components, replacing batteries, and resupplying fuel for generators to remote sites, communications towers may require checkups (to observe such factors as rust and guy-tensioners). Other tasks such as painting and light replacement may also be necessary.

Periodic routine maintenance for telecommunications equipment is performed in accordance with the relevant equipment manuals and manufacturer's requirements. A good *network management system* (NMS) purchased and installed with the communications equipment, will simplify the maintenance tasks and automate the process of periodic maintenance logging and record-keeping.

Maintenance can be performed by the operator, or it can be contracted out to a third-party company. In preparing a maintenance program, one has to define a maintenance system and arrange for the staffing of maintenance personnel; to establish maintenance centers and determine the types and number of test equipment, tools, and spare parts needed for each center; to prepare a maintenance schedule and forms for maintenance logs and records; and to define the target system performance values for providing maintenance and prepare standardized test procedures.

7.7 Testing and Commissioning of the Communications System

After the communications system's equipment, such as fiber-optic cables and microwave links, have been installed, they will have to be tested. A record of the standard field turn-up tests (defined by the operator, agreed upon by the equipment supplier, and performed after the equipment installation is complete) has to be provided, as does the system verification document. The system can be turned on when the as-built documentation is ready and system acceptance tests have been performed.

Commissioning is the process of systematically bringing the various components of the project into an operational mode prior to startup and executing a formal handover to the end user. As soon as the equipment becomes operational and tested according to the ATP, the project manager initiates and submits to the client a project final certificate for acceptance. This certificate normally records the formal handover of the completed project and identifies all the documentation (including operational and maintenance manuals, as-built drawings, equipment warranties, and contract completion reports) that the end user requires for ongoing operation.

There are a number of different tests that may be performed using dozens of pieces of different test equipment. Section 7.7.1 briefly describes only a few of the most important tests: BER, optical fiber and Ethernet.

7.7.1 BER Testing

To measure the BER, enough bits must be sent to allow a meaningful determination of the measurement statistics. Theoretically, as the number of transmitted bits approaches infinity, the BER becomes a perfect estimate of the true error probability.[32] In other words, the probability of receiving a bit in error is exactly equivalent to the BER of a communications link when an infinite number of bits are transmitted and received. Mathematically, we can express this definition as follows:

$$BER = \frac{\text{Number of Errored Bits}}{\text{Total Number of Bits}} = \lim_{n\to\infty}\frac{E(n)}{n} \quad (7.5)$$

where n is the total number of bits transmitted and the E is the number of errored bits.

In practice, we cannot send an infinite number of bits and must accept less than perfect estimation, so we have to determine how many transmitted bits will provide a reasonable *confidence level* (CL) for having achieved a specified BER [17]. Estimations on how many bits have to be transmitted (i.e., how long the test has to be relative to the data rate on the communications system) are widely different in the literature.

With some knowledge of stochastic math and by applying Poisson distribution, which is a special case of the binomial probability distribution, and using an *iterative method*[33] it is possible to calculate the total number of bits required:

$$n = \frac{1}{BER}\left[\ln\left(\sum_{k=0}^{E}\frac{(n\cdot BER)^k}{k!}\right) - \ln(1-CL)\right] \quad (7.6)$$

For $E = 0$, formula simplifies to:

$$n = \left[-\ln(1-CL)\right]\times\frac{1}{BER} \quad (7.7)$$

For the simplest case, where the detected number of errored bits is zero ($E = 0$), and the confidence level is very high, CL = 99%, it is possible to calculate that the total number of transmitted bits, n, should be 4.6 times greater than the reciprocal of the specified BER; in other words, $n = 4.6\times\frac{1}{BER}$.

The test time *(T)* is a product of the required number of bits transmitted, n, and the bit time (which is a reciprocal of the data rate):

$$T = \text{Number of Bits}\times\frac{1}{\text{Data Rate}} = \frac{n}{\text{Data Rate}} \quad (7.8)$$

32. In most cases, it is sufficient to prove that probability of bit error is less than some upper limit and without proving its exact value. For example, the probability of bit error required by many telecommunications systems is 10^{-10} or better.

33. The equation cannot be solved using algebraic methods since variable n appears on both sides of an equation and it cannot be isolated.

It is often the case when measuring low BERs or measuring low data rates that direct measurement of BER on a communications link consumes considerable time and becomes impractical. An alternative approach is to extrapolate the BER from a few data points that require less time to measure.

The basic principle is to deliberately degrade BER by decreasing the SNR to the point where the BER is practical to measure. Then, the sampled BER points are overlaid on a BER-versus-SNR curve and the final link BER is extrapolated from the curve. There are two primary approaches used to degrade the signal to noise ratio—decreasing the signal strength or increasing the noise.

In the field, brief initial tests after the communications circuit is established may involve only sending a few bits and making sure that a test set on the other side can receive them (connectivity test). After that, a few different test patterns can be sent and results recorded over the period of 10 minutes or so.

Long-term measurements are required for radio systems, typically 24 hours or longer, and the results are recorded and analyzed.[34]

7.7.2 Optical Time Domain Reflectometer (OTDR)

7.7.2.1 OTDR Principle of Operation

OTDR operation is based on Rayleigh scattering and Fresnel reflection (Figure 7.4). A small fraction of the light is spread in all directions when it encounters changes in homogeneities of the size of its wavelength. A small part of this light—called backscattered light—is captured by the core of the fiber and propagates backward. It is a local phenomenon, and any variation of the backscattered level along the fiber is due to a defect or to an alteration of the properties at a local point. When measuring, a light pulse is first injected into the fiber(s) under test and then the OTDR measures the amount of backscattered light as a function to the time from the initial pulse [18].

Fresnel reflection occurs when the electromagnetic wave propagates through the border of materials with different dielectric constants. A certain part of the

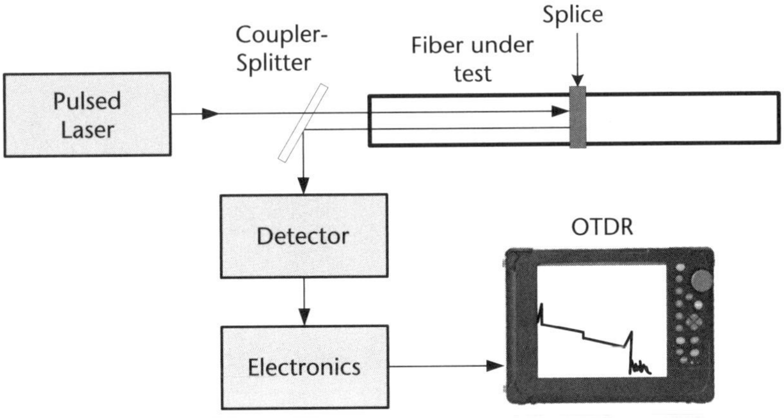

Figure 7.4 OTDR block diagram.

34. Many communications systems, especially microwave radio links could work well during the night and experience problems in early hours of the day due to changes in the radio propagation environment. That is why they have to be tested for longer than just a few minutes.

power is reflected that can be calculated according to Fresnel's law. Such a reflection occurs at the terminations of the fiber and at connectors, where the dielectric constant changes abruptly (e.g., glass-air transition).

Backscattered power depends on the pulse width, the fiber attenuation coefficient, and the *mode field diameter* (MFD). The optical echo is conventionally represented on a logarithmic graph: It therefore appears as a (theoretically) straight line, where the slope represents the attenuation coefficient of the fiber.

The OTDR is preprogrammed with the optical *index of refraction* (IOR) value for the fiber under test to enable the OTDR to calculate and display the length and position of any events (observed as regions of higher or lower levels of reflected or backscattered light) as the measurement pulse travels down the fiber.[35]

OTDRs are always used with a *launch cable* and may use a *receiver cable*. The launch cable allows the OTDR to settle down after the test pulse is sent into the fiber and provides a reference connector for the first connector on the cable under test to determine its loss. A receiver cable may be used on the far end to allow measurements of the connector on the end of the cable under test.

7.7.2.2 OTDR Measurements

During fiber-optic cable installation, the splice loss is estimated by the optical image processing system of the splicer unit. Based on this estimation, the splice can be approved or rejected. When commissioning an optical link, splice losses usually are checked again by OTDR testing from either one side or from both sides of the fiber link.[36]

Figure 7.5 shows some significant events on the OTDR's recorded trace. Event A is a near-end reflection and dead zone; event B usually represents the splice loss (or possibly a bend in the fiber); event C is a reflective event caused by a mechanical splice or crack in the fiber, and event D represents an end of the fiber. Connectors always show both reflection and loss.

Accurate OTDR attenuation measurements depend on having a constant backscatter coefficient. It is commonly assumed that the backscatter coefficient is constant along the fiber length and therefore that the OTDR can be calibrated to read

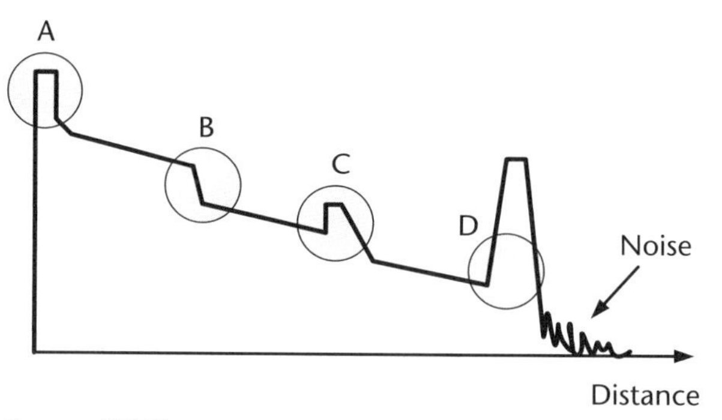

Figure 7.5 Interpreting an OTDR trace.

35. The specific IOR values depend on fiber type and measurement wavelength and they are specified in manufacturer's product information sheets.

36. Note that for testing splices in networks with optical splitters special test procedures have to be followed.

attenuation. Unfortunately, this is often not the case; the backscatter coefficient is, in fact, a function of the core diameter of the fiber (or mode field diameter in a single-mode fiber) and the material composition of the fiber (which determines attenuation). Thus, a fiber with either higher attenuation or larger core size will produce a larger backscatter signal.

The amount of backscattered light is inversely proportional to the MFD and so a fiber with a smaller MFD carries a larger optical power density in the core, and an OTDR detector will see proportionally more backscattered light than a larger MFD fiber. When measuring from a fiber with a larger MFD to one with a smaller MFD, the OTDR measurement will result in a "gainer." Conversely, when measuring from a smaller MFD to a larger MFD, the measurement will result in an *exaggerated loss*. The "true" or actual attenuation or splice loss is calculated from the bidirectional average of the two unidirectional OTDR inspection measurements, which eliminates any spurious backscatter differences from the two unidirectional measurements [19].

The MFD of the fiber is expressed in micrometers (μm). The induced splice loss due to MFD differences is small (typically 0.10–0.20 dB) and can be precisely calculated as:

$$Loss_{dB} = -20\log\frac{2\cdot MFD_1\cdot MFD_2}{MFD_1^2 + MFD_2^2} \tag{7.9}$$

During cable manufacturing, it is necessary to have some excess fiber in the cable to allow the cable to stretch when being pulled without stressing the fiber. This excess fiber is approximately one percent. Since the OTDR measures the length of the fiber, not the cable, it is necessary to subtract one percent from the measured length to get the likely cable length. This becomes important when the OTDR is used to find a fault in an installed cable—this variable adds up to approximately 10m per kilometer; therefore it is not negligible especially on the long systems.

The OTDR has a limited ability to resolve two closely spaced anomalies by the pulse width. In order to enhance the SNR of the received signal, the OTDR sends out many pulses and averages the returned signals. In addition, to get to longer distances, the power in the transmitted pulse is increased by widening the pulse width. This longer pulse width masks all details within the length of the pulse, increasing the minimum distance between features resolvable with the OTDR. As a result, long-distance OTDRs may have a minimum resolution of 250–500m, while short-range OTDRs can resolve features 5–10m apart. This limitation makes it difficult to find problems inside a building where distances are short (a visual fault locator may be required).

It is possible to test multimode fiber with a single-mode OTDR, but not to test a single-mode fiber with a multimode OTDR. When testing multimode with a single-mode unit, the distance measurements will be accurate as long as the correct index of refraction is used, but the loss measurements will appear to be slightly better than they actually are. This is because a splice loss may read as lower in value than it really is because the light from the single-mode OTDR travels mostly in the center of the core, whereas loss occurs at the core edge [20].

7.7.2.3 Testing DWDM Optical Systems

The ITU-T G.694.1 frequency grid specifies the wavelengths used in DWDM. They are found in the C-band (1,525–1,565 nm) and L-band (1,565–1,620 nm), which is a spectral range that is commonly used for DWDM. The reason is because it allows amplification with EDFAs. DWDM networks require specialized testing methods; fibers must be tested for loss, connector cleanliness, dispersion, and spectral quality.

7.7.3 Ethernet Testing

In many cases it is important to test the performance of the communications link end-to-end when the Ethernet links are deployed. A typical Ethernet testing procedure could consist of multiple tests, like port autosensing, jumbo frame support, QoS support, VLAN tagging support, RFC2544 throughput, latency, and burst testing and RFC1242 frame loss rate testing [21].

7.7.3.1 Ethernet Ports and Cables Autosensing

Autosensing is a feature of Ethernet hubs, switches, and NICs and involves probing the capability of the network using low-level signaling techniques to select compatible Ethernet speeds. Autosensing was developed to make the migration from traditional Ethernet to fast Ethernet products easier.

When first connected, devices automatically exchange information with each other to agree on a common speed setting. The devices run at 100 Mbps if the network supports it; otherwise they drop down to 10 Mbps. Many hubs and switches are capable of autosensing on a port-by-port basis, in which case some devices on the network may be communicating at 10 Mbps and others at 100 Mbps. All gigabit Ethernet ports today have autosensing and auto-negotiating capabilities.

An Ethernet crossover cable is used to connect computing devices together directly. Normal straight through or patch cables were used to connect from a host network interface controller (a computer or similar device) to a network switch, hub, or router. A cable with connections that "cross over" or "roll over" was used to connect two devices of the same type, for example, two hosts or two switches to each other. With an *automatic medium-dependent interface crossover* (auto-MDIX) capability modern implementations of the Ethernet over twisted pair standards usually no longer require the use of crossover cables.[37]

7.7.3.2 Jumbo Frames

Jumbo frames are Ethernet frames with 1,500–9,000 bytes of payload.[38] Jumbo frames have the potential to reduce overheads and CPU cycles, but they may also have an adverse effect on network latency, especially on low bandwidth links. An-

37. Auto-MDIX (introduced in 1998) was specified as an optional feature in the 1000BASE-T standard, but today it is implemented widely on most interfaces. Besides the Automatic MDI/MDIX term, this feature may also be referred to by various vendor-specific terms including: Auto Uplink and Trade, Universal Cable Recognition and Auto Sensing, etc.

38. Most gigabit Ethernet switches and gigabit Ethernet network interface cards support jumbo frames. Some fast Ethernet switches and fast Ethernet network interface cards also support jumbo frames.

other important thing to remember here is that the switches, routers, and NICs from one end to the other must all support the same size of jumbo frame transmission for a successful jumbo frame communication session.

To verify that a device supports jumbo frames, it is necessary to configure the *device under test* (DUT) frame size to 4,090 bytes; configure Ethernet test equipment to carry frame sizes of 64, 128, 256, 512, 1,518 and 4,090 bytes; and verify that no frames are dropped.

7.7.3.3 BER Tests in Ethernet Networks

For legacy TDM-based communication systems (T1/E1), the BER tests were traditionally used for installation, commissioning and turn-up, as well as round-trip delay and service-disruption measurements. Unfortunately, there are multiple issues related to using BER testing in Ethernet-based networks. New technologies require new methods and tools for qualification in labs and in the field.

Ethernet is a layer 2 switched technology (data link layer in the OSI seven-layer model). Layer 3 is the network layer; this function is most commonly carried out by IP, but it could be another protocol as well. The purpose of Ethernet is to ensure that data is transferred over a link in a communications network, while the layer 3 protocol has the job of ensuring that the data is transferred over the whole network, from the original source to the ultimate destination, using any number of separate Ethernet links.

The higher layer protocols (layer 4 and above) have the task of ensuring the integrity of transmitted data and presenting the data to the user or application, and as such, are of little interest in a transmission environment.

Considering all this, it becomes obvious that the hardware loopback, commonly used in TDM networks, might not be the best test approach. The integrity of an Ethernet frame is verified at each switching element, and a single bit error will result in the *frame check sequence* (FCS) error and result in the entire frame being discarded. The errored bit will never get to the analyzer, and the analyzer will declare a frame loss. For this reason, BER, the most fundamental measure of a SONET/SDH and older PDH service, has no meaning in Ethernet since the ratio of good to errored bits cannot be defined.

For pure link-to-link Ethernet connections, or even in metro Ethernet, a BER test is still useful whenever the traffic does not interface nodes that can drop frames.

7.7.3.4 IETF RFC 2544

Many Ethernet and GigE test solutions have adopted the guidelines of the Internet Engineering Task Force recommendation RFC 2544, which was originally designed for verifying the performance of LAN devices at the MAC layer. RFC 2544 describes a set of procedures for measuring the performance of Ethernet equipment but it is also used for the overall network.

For Ethernet-based backhaul services, the tests generally used are throughput and latency/frame delay (according to RFC 2544), as well as packet delay variation (according to RFC 3393), but some additional tests are required to validate the correct operation. Measuring frame loss over time will provide an idea of the long-term integrity of the service.

The standard sizes for Ethernet testing are 64; 128; 256; 512; 1,024; 1,280; and 518 bytes. For systems that support jumbo frames, these frame sizes should be tested as well. With the exception of the reset test, each of the tests is run separately for each specified frame size. Small frame sizes increase the number of frames transmitted, thereby stressing the network device as it must switch a large number of frames.

The tests can take from a few minutes to a few hours depending on the throughput of the network and the test parameters (e.g., duration of each test, frame sizes, and resolution). As an intrusive test, RFC 2544 cannot be performed in a monitoring mode. Parameters that are analyzed during the Ethernet acceptance testing:

- Throughput: The throughput test defines the maximum number of frames per second that can be transmitted without any error. The test is done to measure the rate-limiting capability of an Ethernet switch as found in carrier Ethernet services [22], and it must be performed for each frame size. Although the test time during which frames are transmitted can be short, it must be at least 60 seconds for the final validation. Each throughput test result must then be recorded in a report, using frames per second (fps) or bits per second (bps) as the measurement unit.
- Latency (delay): Latency is a key performance metric and is defined as the elapse of time between the request for data and its delivery. A number of factors contribute to Ethernet message delay: store and forward, switch fabric, transmission medium (wireline or radio), and queuing delays. The implementation and usage details of a network, including the number and layout of switches (both physical and logical), the number and distribution of receivers and transmitters, their connection speeds and traffic patterns, all contribute to the total end-to-end system delay.

  The latency test (also known as end-to-end testing) measures the time required for a frame to travel from the originating device through the network to the destination device. The total delay time is the sum of both the processing delays in the network elements and the propagation delay along the transmission medium. Delays caused by Ethernet switch electronic circuits are known as the switch fabric delay, and typical values are 5–10 μs.

  In LANs, transmission media delay (fiber-optic, copper, microwave radio) is negligible in comparison to other contributions to the total end-to-end delay. Only when very long-distance Ethernet links are deployed can this delay become significant.
- Frame loss rate: The frame loss test measures the network's response in overload conditions, a critical indicator of the network's ability to support real-time applications in which a large amount of frame loss will rapidly degrade service quality. As there is no retransmission in real-time applications, these services might quickly become unusable if frame loss is not controlled.
- Back-to-back frames: The back-to-back frames test determines the maximum number of frames sent back-to-back with minimal interpacket gap (IPG), or in other words, at 100% frame rate, which the DUT can process without losing frames. The test begins with a specified number of frames and repeats with more or fewer frames until it determines the maximum number. As

always, this is repeated for each frame size. This test may go by other names, such as the burstability or maximum throughput test.

- System recovery: The system recovery speed is the time the DUT takes to stop losing frames when the frame rate is reduced from a stressed to a normal state. The device is stressed for at least one minute with too much traffic, typically 110% of the rate determined by the throughput test, such that it loses frames. Then, the traffic is reduced to 50% of the predetermined throughput. The time between the drop in frame rate to the last lost frame is averaged over several test runs for each frame size.
- Reset: The reset test measures the time it takes for the DUT to begin forwarding frames after hardware or software reset or a power interruption. Unlike the other RFC 2544 tests, only the smallest frame size, typically 64 bytes, needs to be tested. The delay between the last frame forwarded before the reset and the first frame after the reset is the reported reset speed. The test should be run for each type of reset.
- Oversubscription: The most common reason for frame loss is oversubscription of the available bandwidth. For example, if two 1,000-Mbps Ethernet services are mapped into a single 622-Mbps SONET/SDH pipe (a common scenario), the bandwidth limit is quickly reached, and as a result, frames may be dropped.

  Some network elements are able to prioritize traffic based upon VLAN ID or priority tag. If this function is being used, as the bandwidth limit is reached it should be the low-priority packets that get dropped, and this functionality has to be tested as well.
- Packet delay variation (PDV): This measurement is critical because excessive PDV can cause major failures in real-time applications such as VoIP or streaming video. It is also very important for teleprotection in electric utility environment.

  At the start of the transmission, the interframe gap between all frames was identical. As the frames navigate through the network, being buffered or routed thought different network devices, the interframe gap begins to vary. The PDV measurement should be done at maximum frame rate as this is where the most variation will occur.

  The variation in packet delay is sometimes called *jitter*. This term could be confusing because it is understood differently by different groups of people. "Jitter" commonly has two meanings: The first meaning is the variation of a signal with respect to some clock signal, where the arrival time of the signal is expected to coincide with the arrival of the clock signal. This meaning is used with reference to synchronous signals and might be used to measure the quality of circuit emulation, for example. There is also a metric called *wander* used in this context.

  The second meaning has to do with the variation of a metric (e.g., delay) with respect to some reference metric (e.g., average delay or minimum delay). This meaning is frequently used by computer scientists and frequently, although not always, refers to variation in delay.

- QoS: Layer 2 Ethernet switches rely on IEEE 802.1p standard to provide QoS. The standard 802.1p is the part of the IEEE 802.1Q (2005) that defines the architecture of virtual bridged LANs (VLANs) [23]. If the system requires (and supports) QoS, it is important to double-check if the device has traffic priorities set for 802.1p, and using Ethernet test equipment, to verify that the traffic has no dropped frames.
- VLAN tagging: A single layer-2 network may be partitioned to create multiple distinct and mutually isolated broadcast domains, referred to as a VLAN. This is usually achieved on switch or router devices. Simpler devices only support partitioning on a port level (if at all), so sharing VLANs across devices requires running dedicated cabling for each VLAN. More sophisticated devices can mark packets through tagging, so that a single interconnect (trunk) may be used to transport data for various VLANs. IEEE's 802.1Q VLAN tagging is the industry standard.

  VLAN membership can be configured through software instead of physically relocating devices or connections. Unlike physically separate networks, VLANs share bandwidth, so VLAN trunks may require aggregated links and/or QoS prioritization.

7.7.3.5 ITU-T Y.1564

When completing all tests based on recommendation IETF RFC 2544 at all frame sizes, the required time could be long. Due to this reason, often service providers complete only a subset of the tests.

With networks evolving, the testing standard has to address different types of network configurations and ensure that quality is maintained across networks with multiple streams with different policing parameters. Recommendation *ITU-T Y.1564* defines an out-of-service test methodology that may be used in assessing the proper configuration and performance of an Ethernet network to deliver Ethernet-based services [24].

In order to simply the test and address all areas, the standard has been written around two core components—the *Service Configuration Test* and the *Service Performance Test.* During the Service Configuration Test each of the individual test flows are completed in a sequentially manner, confirming there are no network configuration issues. Once the network configuration test has been completed, it is necessary to confirm that the network is able to work under load over a configured duration of time.

The Service Performance Test completes this phase by generating all test flows simultaneously at the committed information rate (CIR) for a period of time of 15 minutes, two hours or 24 hours (based on ITU-TM.2110), or user-selectable [25].

Placing the network under load with multiple test flows for this duration allows the engineer to see if other services are affecting the service under test but also allows for statistics such as availability and unavailability to be measured.

7.8 Review Questions and Problems

1. What are the main design parameters when it comes to selection of the appropriate communications tower relative to its function, loading, and location?
2. In what situations is it necessary to use plenum-rated communications cables? Explain briefly your rationale.
3. A 75-ohm load is connected to a 100-ft-long, low-loss transmission line with a characteristic impedance of 50 ohms. What is the VSWR at the termination?

   Answer:

   For a lossy transmission line, the VSWR is a function of the length of the line as well as the load impedance and the characteristic impedance of the line. However, for a lossless (or low-loss) transmission line, the VSWR is the same at any reference point of the line. That means that the length of the transmission line does not have any significance in VSWR calculation.

   The reflection coefficient, Γ, is calculated as:

$$\Gamma = \frac{Z_L - Z_0}{Z_L + Z_0} = \frac{75-50}{75+50} = \frac{25}{125} = 0.2$$

   The return loss is:

$$R_L = -20\log\Gamma = -20\log 0.2 \approx 14\,\mathrm{dB}$$

   and the VSWR is:

$$VSWR = \frac{1+\Gamma}{1-\Gamma} = \frac{1+0.2}{1-0.2} = \frac{1.2}{0.8} = 1.5$$

   Discussion:

   The VSWR is less than two and can be considered a very good match.
4. Explain the OTDR principle of operation. When are bidirectional measurements required when performing tests with OTDR and why?
5. Find out how long the bench test of the RBER[39] of a 155-Mbps microwave radio (in a back-to-back configuration) has to be in order to confirm with 99% confidence a value of RBER = 10^{-12} or better. RBER measurement in a lab environment does not require any Rayleigh fading (radio propagation) simulation since it represents the quality of the microwave radio hardware itself.

39. Residual BER characterizes the radio in its normal operating received signal strength range. It measures the combined effect of the digital radio's modulator, transmitter, receiver, and demodulator. Raw uncorrected BER rates of 10^{-10} are usually sufficient to achieve 10^{-12} after correction (FEC – Forward Error Correction). A radio "dribbling" errors (small numbers of errors scattered across time) at normal received signal levels is a classic sign of a residual BER problem.

Answer:

In order to have 99% confidence that the RBER of the radio system is 10^{-12} or better, we have to send $n = 4.5 \times \frac{1}{BER} = 4.6 \times \frac{1}{10^{-12}} = 4.6 \times 10^{12}$ bits, and receive zero errors within the test period. Once we have tested this many bits without error, we can be sure that our actual BER is better than 10^{-12}.

$$\text{Time} = \frac{n}{155\ \text{Mbps}} = \frac{4.6 \times 10^{12}\,\text{bits}}{1.55 \times 10^{8}\ \text{bps}} = 8.24\ \text{hours}$$

Discussion:

This measurement assumes random Gaussian or normal distribution and will take approximately eight hours. If the error distribution is not Gaussian, an even longer period of measurement is needed to validate the RBER measurement.

A. Test time can be reduced but at the expense of a reduced confidence level in the test results. Calculate the number of bits and the test time required for a confidence level of 95%. *(Answer:* $n = 3 \times \frac{1}{BER}$ and the test would require 5.4 hours).

6. List the main functions of a project manager on a large communications project. What skill set is expected from a project manager?
7. At what point on every telecommunications project does it become necessary to discuss and define system maintenance?
   A. Who are the main stakeholders involved in this decision and why?
   B. Describe the maintenance required on different types of radio and fiber-optic systems.
   C. What is your understanding of the term *lifetime warranty* on a product? Is there an actual number of years attached to this term?
   D. Could unmanned aerial systems (UASs), more commonly known as drones, be used to inspect communications towers and equipment installed on those towers? What are the FAA requirements to fly commercially deployed drones? Discuss FAA Section 333 [26].
8. Please explain why PDV is critical in packet networks for real-time applications such as VoIP, streaming video, and teleprotection in an electric utility environment?
9. What is a demarcation point, and why it is commonly used to troubleshoot communications network problems? Describe types of demarcation points.
10. GPS is a satellite system, used in the United States for different applications. Summarize and show in a table format main features and applications of GPS and compare it with other satellite systems with a similar purpose (GLONASS, Compass, IRNSS, and Galileo).

References

[1] Lehpamer, H., *Transmission Systems Design Handbook for Wireless Networks*, Norwood, MA: Artech House, 2002.

[2] National Wildfire Coordinating Group, National Interagency Incident Management System: "Basic Land Navigation," NFES 2865, June 2007.

[3] http://oceanservice.noaa.gov/facts/lidar.html (accessed October 13, 2015).

[4] Kaplan, E. D., ed., *Understanding GPS: Principles and Applications,* Norwood, MA: Artech House, 1996.

[5] IEEE Std 525-2007, IEEE Guide for the Design and Installation of Cable Systems in Substations, 2007.

[6] Transport Canada, *Standard 621— Obstruction Marking and Lighting,* First Edition, 2011.

[7] EG&G Services, Parsons, Inc., Science Applications International Corporation, *Fuel Cell Handbook,* Fifth Edition, October 2000.

[8] http://eosweb.larc.nasa.gov (accessed January 26, 2015).

[9] Wysocki, R. K., et al., *Effective Project Management,* New York: John Wiley and Sons, 1995

[10] Kerzner, H., *Project Management—A Systems Approach to Planning, Scheduling, and Controlling,* Seventh Edition, New York: John Wiley and Sons, 2001.

[11] Clark, M. P., *Networks and Telecommunications—Design and Operations,* Second Edition, New York: John Wiley & Sons, 1997.

[12] Situation Management Systems, Inc., *Managing Negotiation—Selected Readings on Negotiation Skills,* 1996.

[13] http://www.weather.gov (accessed Oct 18, 2013).

[14] Cho, K. T., *Multicriteria Decision Methods: An Attempt To Evaluate and Unify,* School of Systems Management Engineering, Sungkyunkwan University, Suwon, Korea, 1995.

[15] Saaty, T. L., *The Analytic Hierarchy Process,* NY, McGraw-Hill, 1980.

[16] Fülöp, J., *Introduction to Decision Making Methods,* Laboratory of Operations Research and Decision Systems, Computer and Automation Institute, Hungarian Academy of Sciences, 2005.

[17] Technical Article, *Physical Layer Performance: Testing the Bit Error Ratio (BER)*, HFTA-010.0, Maxim Integrated, 10/2004.

[18] TIA/EIA-455-8, *Measurement of Splice or Connector Loss and Reflectance Using an OTDR*, 2000.

[19] Corning, *AN3060: Guidance for OTDR Assessment of Fusion Spliced Single-mode Fibers,* Application Note, March 2014.

[20] Anritsu Corporation, *Understanding OTDRs,* Issue 1, 2011.

[21] *IETF RFC 2544 "Benchmarking Methodology for Network Interconnect Devices,"* 1999.

[22] Giguere, B., "RFC 2544: How it Helps Qualify a Carrier Ethernet Network," Application Note 183, *EXFO, Canada,* 2004.

[23] IEEE Std 802.1Q, *IEEE Standard for Local and Metropolitan Area Networks—Virtual Bridged Local Area Networks,* 2005.

[24] *ITU-T Y.1564, Ethernet Service Activation Test Methodology,* 2011.

[25] *ITU-T M.2110, Bringing into Service International Multi-Operators Paths, Sections and Transmission Systems,* 2002.

[26] https://www.faa.gov/uas/legislative_programs/section_333/ (accessed October 14, 2015).

Appendix A
Review of Engineering Math

In order to be able to follow the material in this book, one does not require knowledge of calculus but should have a very good understanding of algebra, especially powers and exponents, logarithms, and exponential and logarithmic equations. This is just a brief refresher; a more in-depth review can be found in other sources [1, 2].

A.1 Powers and Exponents

A.1.1 Basic Rules of Powers and Exponents

$$A^m \cdot A^n = A^{m+n} \qquad \frac{A^m}{A^n} = A^{m-n}$$

$$\left(A^m\right)^n = A^{m \cdot n} \qquad (A \cdot B)^n = A^n \cdot B^n \qquad A^{-m} = \frac{1}{A^m}$$

$$A^{1/n} = \sqrt[n]{A} \qquad \sqrt[n]{A \cdot B} = \sqrt[n]{A} \cdot \sqrt[n]{B} = A^{1/n} \cdot B^{1/n}$$

$$\sqrt[n]{\frac{A}{B}} = \frac{\sqrt[n]{A}}{\sqrt[n]{B}} = \frac{A^{1/n}}{B^{1/n}} \qquad A^{m/n} = \sqrt[n]{A^m} \qquad \sqrt[m]{\sqrt[n]{A^m}} = A^{1/m \cdot n}$$

A.1.2 Scientific Notation

Scientific notation is a means of compactly writing very large or very small numbers and can be useful for making order or magnitude estimates. By definition, a number in scientific notation is a product of a number between one and 10 and an integer power of 10.

$$1,250,000 = 1.25 \times 10^6$$
$$0.000071 = 7.1 \times 10^{-5}$$

$$\left(4\times10^{8}\right)\times\left(5\times10^{-5}\right)\times\frac{2\times10^{3}}{8\times10^{4}}=$$
$$=\left(4\times10^{8}\right)\times\left(5\times10^{-5}\right)\times\left(2.5\times10^{-2}\right)=$$
$$=\left(4\times5\times2.5\right)\times10^{8+(-5)+(-2)}=$$
$$=5\times10^{2}$$

A.2 Logarithms

A.2.1 Definition of a Logarithm

The logarithm of A is the power to which base X would have to be raised to equal A.

$$A=X^{B}\Rightarrow B=\log_{X}A$$

X is the base of a logarithm, and base 10 (also called decimal or Briggsian) is most commonly used in communications. Base 2 (also called *binary*) or base of natural (also called *Neperian*) logarithms $e = 2.71828$ are used sometimes as well. The notations $ln\ A$ and $log_e\ A$ both refer unambiguously to the natural logarithm of A.

A.2.2 Basic Rules of Logarithms

$$\log_{X}A\cdot B=\log_{X}A+\log_{X}B$$
$$\log_{X}\frac{A}{B}=\log_{X}A-\log_{X}B$$
$$\log_{X}A^{n}=n\cdot\log_{X}A$$
$$\log_{X}\sqrt[n]{A}=\frac{1}{n}\cdot\log_{X}A$$
$$\log_{X}1=0$$
$$\log_{X}X=1$$

A.2.2.1 Example

$$10\log\left(\frac{A^{2}}{B}\times C\right)=10\log A^{2}+10\log C-10\log B=$$
$$=20\log A+10\log C-10\log B$$

The equation for changing the base of the logarithm from base X to base Y is:

$$\log_{Y}A=\frac{\log_{X}A}{\log_{X}Y}$$

The logarithm of negative numbers can be defined only among the complex numbers.

A.2.3 Exponential and Logarithmic Equations

Most exponential equations do not solve neatly like in those Examples 1 and 2; in solving these more complicated equations (Examples 3–7), we need to use logarithms. To solve an exponential equation, take the logarithm of both sides, and solve for the variable. A logarithmic equation is one in which a logarithm of the variable occurs (Example 4).

Examples 7 and 8 show that the same rules apply to all logarithms, regardless of the base.

Example 1 $2^x = \frac{1}{64}$ $2^x = 2^{-6}$ $x = -6$	Example 2 $9^{2x-5} = 27$ $\left((3)^2\right)^{2x-5} = 3^3$ $3^{2(2x-5)} = 3^3$ $4x - 10 = 3$ $4x = 13$ $x = \frac{13}{4}$ $x = 3.25$	Example 3 $2^x = 100$ $\log 2^x = \log 100$ $x \cdot \log 2 = \log 100$ $x = \frac{\log 100}{\log 2} = \frac{2}{\log 2}$ $x \approx 6.644$
Example 4 $x^5 = 1000$ $\log x^5 = \log 1000$ $5 \cdot \log x = \log 1000$ $\log x = \frac{\log 1000}{5}$ $x = 10^{\frac{\log 1000}{5}} = 10^{\frac{3}{5}} = \sqrt[5]{10^3}$ $x \approx 3.981$	Example 5 $\log_2(x+12) = 6$ $x + 12 = 2^6$ $x + 12 = 64$ $x = 52$	Example 6 $5^{2x-1} = 4$ $\log 5^{2x-1} = \log 4$ $(2x-1)\log 5 = \log 4$ $2x \log 5 - \log 5 = \log 4$ $2x \log 5 = \log 4 + \log 5$ $x = \frac{\log 4 + \log 5}{2\log 5}$ $x \approx 0.9307$
Example 7 $x = 10\log_2 1000$ $\frac{x}{10} = \log 1000$ $2^{\frac{x}{10}} = 1000$ $\log 2^{\frac{x}{10}} = \log 1000$ $\frac{x}{10}\log 2 = \log 1000$ $x = \frac{10\log 1000}{\log 2} = \frac{10 \cdot 3}{\log 2}$ $x = \frac{30}{\log 2}$ $x \approx 99.66$	Example 8 $x + 5 = \log_5 10$ $5^{x+5} = 10$ $(x+5)\log 5 = \log 10$ $x + 5 = \frac{\log 10}{\log 5}$ $x = \frac{\log 10}{\log 5} - 5$ $x = \frac{1}{\log 5} - 5$ $x \approx -3.57$	

References

[1] Fanchi, J. R., *Math Refresher for Scientists and Engineers* (Third Edition), Hoboken, New Jersey: John Wiley and Sons, Inc., 2006.

[2] Bronshtein, I. N., et al., *Handbook of Mathematics* (Fifth Edition), Berlin and Heidelberg, Germany: Springer-Verlag, 2007.

Appendix B
Physical Quantities and Units of Measurements

North Americans probably use a greater variety of units of measurement than anyone else in the world. Not too many people know that the legal definitions of the English customary units are actually based on metric units. The U.S. and British governments have agreed that a yard equals exactly 0.9144m and that an avoirdupois pound equals exactly 0.45359237 kg. In this way, all the units of measurement North Americans use every day are based on the standards of the metric system.

Since 1875, in fact, the United States has subscribed to the International System of Weights and Measures, the official version of the metric system.[1] All systems of weights and measures, metric and nonmetric, are linked through a network of international agreements supporting the International System of Units, called the SI.[2] At the heart of the SI is a short list of base units defined in an absolute way without referring to any other units. The base units are consistent with the part of the metric system called the MKS system.

In the metric system it is very easy to create a larger (or smaller) unit of measurement simply by adding a prefix in front of base unit.

B.1 Physical Constants

c = speed of light in vacuum

$$c = 299{,}792{,}458 \text{m}/s \approx 3 \cdot 10^8 \text{ m/sec}$$

or

$$c = 9.835711 \cdot 10^8\ ft/\text{sec}$$

1. The key agreement is the Treaty of the Meter (Convention du Mètre), signed in Paris on May 20, 1875. Forty-eight nations have signed this treaty, including all the major industrialized countries.

2. It comes from the first two initials of its French name, Système International d'Unités (ITU-R Recommendation V.430-3: "Use of the International System of Units (SI)").

B.2 Mathematical Constants

e = 2.7182818 base of natural logarithms

Frequency:

1 GHz = 10^3 MHz = 10^6 kHz = 10^9 Hz

Distance:

1 mile[3] = 1.609 km = 5,280 ft = 63,360 in[4]
1m = 1.09361 yd = 3.28084 ft = 0.001 km = 6.21371×10^{-4} mi = 39.3701 in
1 ft = 12 in = 0.3048 m = 0.3333 yd
1 chain = 66 feet
1 mile = 80 chains
1 nautical mile (1 NM) = 1.151 mile

Area:

1 m^2 = 1,550 in^2 = 10.7639 ft^2 = 1.19599 yd^2
1 acre[5] = 208 ft × 208 ft = 43,560 ft^2
1 acre = 10 square chains
640 acres = 1 square mile

Volume:

1m^3 = 61,023.7 in^3 = 1.30795 yd^3 = 35.3147 ft^3

Speed:

1 km/hr = 0.277778 m/s = 0.621371 mi/hr = 3.28084 ft/s
1 mi/hr = 1.609344 km/hr

Mass:

1 lb = 0.453592 kg
1kg = 2.20462 lb

3. The word "mile" comes from the Latin term "mille passum," literally translated as "thousand paces," a unit introduced to Britain during the Roman occupation (57 BCE–450 CE). Each Roman pace is equal to five (5) Roman feet; hence there were 5,000 Roman feet in a Roman mile. But why does the British mile has grown to 5,280 feet? Great Britain has another unit - "furlong" - for measuring length and area. A furlong has 660 ft and it does not divide into 5,000 evenly. So Elizabeth I defined a mile to be 8 furlongs and hence 5,280 feet. As a side note, a furlong is the distance a team of oxen could plow without needing a rest.

4. 1 inch is defined to be exactly 2.54 cm (or 25.4 mm) in July, 1959. Before this, the U.K. inch measures 2.53998 cm, while the U.S. inch was 2.540005 cm.

5. An acre is originally defined as the amount of land a pair of oxen could plow in a day (and we are still using it today?).

Force:

$$1\text{N} = 0.10197 \text{ kp} = 0.2248 \text{ lbf}$$

Pressure:

$$1 \text{ N/m}^2 = 10^{-5} \text{ bar} = 0.02088 \text{ lbf/ft}^2 = 0.10197 \text{ kp/m}^2 = 9.86923 \times 10^{-6} \text{ atm}$$

Energy:

$$1 \text{ kWh} = 3.6 \times 10^6 \text{ J} = 859.845 \text{ kcal} = 3412.14 \text{ BTU}$$
$$1 \text{ kcal} = 4186.8 \text{ J} = 0.745700 \text{ kWhr}$$

Power:

$$1 \text{ hp} = 76.0402 \text{ kpm/s} = 745.700 \text{ W}$$
$$1 \text{ W} = 0.238846 \text{ cal/s} = 3.41214 \text{ BTU/hr}$$

Temperature:

$$°C = \frac{5}{9}(°F - 32)$$

$$°F = \frac{9}{5}(°C) + 32$$

Throughput:

$$1 \text{ byte (or octet)} = 8 \text{ bit} = 2 \text{ nibble}$$
$$1 \text{ GB} = 10^3 \text{ MB} = 10^6 \text{ kB} = 10^9 \text{ byte}$$

Angle:

$$1 \text{ mrad (milliradian)} = 0.0572958°$$
$$1° = 17.45329 \text{ mrad}$$
$$1° = 60' = 3{,}600''$$

B.3 American Cable Stranding

When working on international projects, the information shown here will come in handy. Most of the world expresses wire size in terms of its diameter, but in North America, *American wire gauge* (AWG) is the most common unit. AWG of the wire can be calculated if the area of the wire is known:

$$AWG = -4.3125 \ln\left(A_{mm^2}\right) + 17.16$$

$$A_{mm^2} = e^{\frac{AWG - 17.16}{-4.3125}}$$

AWG of the wire can be calculated if the diameter of the wire is known:

$$AWG = -8.625\ln(d_{mm}) + 18.20175$$

$$d_{mm} = e^{\frac{AWG-18.20175}{-8.625}}$$

1 mil = 1/1000 in = 0.0254 mm
1 CM = 1 circular mil[6] = 0.0005067 mm^2
1 MCM (or sometimes used kcmil) = 1,000 circular mils = 0.5067 mm^2
4/0 is also known as 0000

A Few Rules of Thumb

- When the diameter of a wire is doubled, the AWG will decrease by 6;
- When the cross-sectional area of a wire is doubled, the AWG will decrease by 3.

6. 1 CM is an area of the circle that has 1/1000 inch diameter.

Appendix C
Formulas Reference Sheet

C.1 Frequency and Wavelength

Wavelength can be calculated as:

$$\lambda = \frac{c}{f}$$

where

f = frequency [Hz];

c = speed of light in vacuum. ($c \approx 3 \cdot 10^8$ m/sec or 186,000 mi/sec)

C.2 Signal-to-Noise Ratio

The signal-to-noise ratio (S/N) is the amount by which a signal level exceeds the noise level:

$$\text{S/N [dB]} = \text{signal level [dBm]} - \text{noise level [dBm]}$$

C.3 EIRP

Effective isotropic radiated power (EIRP) describes the performance of a transmitting system:

$$\begin{aligned} &\text{EIRP}\left[\text{dBW or dBm}\right] = \text{Tx output power}\left[\text{dBW or dBm}\right] \\ &+\text{antenna gain}\left[\text{dBi}\right] - \text{line loss}\left[\text{dB}\right] \end{aligned}$$

C.4 Fade Margin

Fade margin (FM) is an "extra" signal power added to a link to ensure its continued operation if it suffers from signal propagation effects.

$$\text{FM}[\text{dB}] = \text{system gain} + \text{ant. gain}(\text{Tx and Rx}) - \text{propagation losses}$$
$$-\text{cable/connector/branching loss } (\text{both ends added together})$$

C.5 System Gain

System gain is the total gain of the radio system without considering antennas and cables. Tx power and Rx threshold sensitivity are expressed in dBm.

$$\text{System gain} = \text{Tx power} - \text{Rx}(\text{threshold}) \text{ sensitivity } [\text{dB}]$$

C.6 Free-Space Loss

$$\text{FSPL} = 96.6 + 20\log_{10}(d_{miles}) + 20\log_{10}(f_{GHz}) \ [\text{dB}]$$
$$\text{FSPL} = 92.4 + 20\log_{10}(d_{km}) + 20\log_{10}(f_{GHz}) \ [\text{dB}]$$

C.7 Parabolic Antenna Directive Gain

$$G_a = \log \eta \frac{4\pi A_a}{\lambda^2} \ [\text{dBi}]$$

where

G_a = antenna directive gain

η = aperture efficiency (usually 55%)

A_a = antenna aperture area

λ = wavelength

Glossary

ABAM A designation for 22-gauge, 100-Ω, plastic insulated, twisted pair Western electric cable, normally used in a CO for T1 circuits. ABAM cable is no longer available, but it is easy to find a replacement cable that meets the technical requirements.

ADC Analog-to-digital conversion. The ADC converts analog signals (for example, signals coming from the sensors) into a digital form.

ADM Add/drop multiplexer. In an ADM, traffic can be dropped or added, or it can simply go through it.

ADSS All-dielectric self-supporting fiber-optic cable; it is ideal for installation in electrical transmission and distribution environments.

AES Advanced encryption standard. AES is defined in FIPS 197 by NIST.

ALBO Automatic line buildout. Electronic circuits used in the receive equipment of the digital transmission system to establish the proper signal level and pulse shape required by the receiver.

AMI Advanced meter infrastructure. Electricity meters that use two-way communications to collect electricity usage and related information from customers and to deliver information to customers.

AMR Automatic meter reading.

ANSI American National Standards Institute.

Antenna beamwidth The directivity of a directional antenna. Defined as the angle between two half-power (–3-dB) points on either side of the main lobe of radiation.

ASL Above sea level.

ASR Antenna structure registration. The Antenna Structure Registration Program is the process under which each antenna structure that requires FAA notification, including new and existing structures, must be registered with the FCC by its owner.

ASTM International American Society for Testing and Materials International.

Asynchronous A type of transmission in which each character is transmitted independently and without reference to a standard clock.

ATM Asynchronous transfer mode. ATM is a packet-oriented digital data transmission technology.

ATPC Adaptive transmit power control. ATPC is a feedback control system that increases the transmitting power level during periods of fading.

Attenuation Reduction in signal magnitude or signal loss, usually expressed in decibels.

Auto-MDIX *See* MDIX.

AWG American wire gauge. A standard system used in the United States for designating the size of an electrical conductor based on a geometric progression between two conductor sizes. Copper phone wiring usually comes in 24- or 26-AWG.
Azimuth The angle between true north and the projection of a surface normal to the horizontal plane, measured clockwise from the north.
Backhaul The portion of the wireless network that carries the wireless calls from cell site radios back to the mobile switching center (MSC) and then on to the appropriate service termination points, such as the PSTN and/or the Internet.
Bandwidth The information capacity of a communications resource, usually measured in bits per second for digital transmission and hertz for analog transmission.
Bellcore Bell Communications Research (*see* Telcordia).
BER Bit error ratio, also referred to as "bit error rate." The number of erroneous bits received divided by the total number of bits transmitted.
BERT Bit error ratio test. Testing method for digital communications circuits and links that uses predetermined stress patterns comprising of a sequence of logical ones and zeros (e.g., QRSS, 3 in 24, all Ones, and all zeros).
BLSR Bidirectional line-switched ring.
Bits/s/Hz Bits per second per Hertz; a measure of spectral efficiency of the radio system.
BLE Bluetooth low energy. It was adopted in 2010 and it includes classic Bluetooth, Bluetooth high-speed and Bluetooth low energy (BLE) protocols.
BLM Bureau of Land Management. An agency within the United States Department of the Interior that administers America's public lands. Most public lands are located in western states, especially Alaska.
BoM Bill of materials; also called bill of quantities (BoQ).
Bonding The permanent joining of metallic parts to form an electrically conductive path that will ensure electrical continuity and the capacity to conduct any current to be present in a safe manner.
BoQ Bill of quantity; also called BoM.
bps Bits per second; number of bits in one second.
Broadband A classification of the information capacity (bandwidth) of a communications channel. Broadband is generally taken to mean a bandwidth higher than 2 Mbps.
Call Sign A combination of letters and numbers that identify an FCC license.
Canadian Electrical Code (CEC) Canadian version of the U.S. National Electrical Code (NEC).
CAPEX Capital expenditures.
Carrier A telecommunications provider that owns switch equipment.
Carrier Ethernet A standardized carrier-class service defined by five attributes that distinguish carrier Ethernet from LAN-based Ethernet: standardized services, scalability, reliability, QoS, and service management.
CCIR International Radio Consultative Committee (now ITU-R).
CE Conformité Européenne (the European EMC standard).
CEPT Conférence Européenne des Postes et Télécommunications (European Conference of Postal and Telecommunications Administrations).
C/I Carrier-to-interference ratio; typically expressed in decibels.
CIP Critical infrastructure protection.

Circuit switching The basic switching process whereby a circuit between two users is opened on demand and maintained for their exclusive use for the duration of the transmission, as opposed to a dedicated circuit that is held open regardless of whether data is being sent.

Characteristic impedance The ratio of voltage to current at any given point on a transmission line.

CLEC Competitive local exchange carrier. Deregulated local telephone companies resulting from the Telecommunications Act of 1996 that are competing for local exchange service, as well as for long distance and Internet service.

CO Central office. The building that contains the switches for a local telephone company.

COFDM Coded orthogonal frequency division multiplex.

COFETEL Comisión Federal de Telecomunicaciones. Established in 1996, the primary agency responsible for communications in the Mexico, an autonomous administrative and technical agency, which resides within the Secretaria de Comunicaciones y Transportes (SCT), a cabinet-level position whose secretary is appointed by the Mexican president.

Combiner An RF device used to combine two or more radio signals to be transmitted via the same antenna, while maintaining the characteristic impedance of the system. Dependent upon the type of combiner, it may introduce additional loss by using resistors, or it may use transformers in which case it could theoretically be lossless.

Conduit *See* duct.

CONUS Continental United States. It refers to the 48 U.S. states located on the North American continent south of the U.S. border with Canada, plus the District of Columbia. The term excludes the states of Alaska and Hawaii and all off-shore U.S. territories and possessions, such as Puerto Rico.

Corona Discharge An electrical discharge at the surface of a conductor accompanied by the ionization of the surrounding atmosphere. It is normally accompanied by light and audible noise.

COTS Commercial off the shelf. Standard systems that are readily available for purchase.

CPE Customer premises equipment.

CRC-n Cyclic redundancy check-n bits. A CRC is an error-detecting code, invented in 1961.

CRTC Canadian Radio and Telecommunications Commission.

Cross-connect Also known as DSX Panel or MDF (main distribution frame). Distribution system equipment used to terminate and administer communications circuits. In a wire cross-connect, jumper wires or patch cords are used to make circuit connections. In an optical cross-connect, fiber patch cords are used. The cross-connect is located in an equipment room, riser closet, or satellite closet.

CSU Channel service unit; the interface from CPE to the public T1 line.

CTI Craft terminal interface; a simple text-based interface, available on most microwave radios for radio operation, administration and maintenance from a VT100-compatible terminal or from a laptop.

DA Distribution automation. A system consisting of line equipment, communications infrastructure, and information technology that is used to gather intelligence about an electrical distribution system. It provides analysis and control in order to optimize operating efficiency and reliability.

DACS Digital access and cross-connect system.

Dark Fiber Dark fiber refers to unused fibers, available for use.

DB-9 A standardized connector with nine pins.

dBm Decibels relative to 1 mW.

dBW Decibels relative to 1W.

DEM Digital elevation model. A digital model describing the elevation of a map area.

DFS Dynamic frequency selection. A mechanism to allow license-exempt devices to share a spectrum with existing military radar and weather radar systems. DFS allows automatic detection of radar signal interference and dynamically switches to a clear channel.

DGPS Differential global positioning system. DGPS is an enhancement to global positioning system that uses a network of fixed, ground-based reference stations to broadcast the difference between the positions indicated by the satellite systems and the known fixed positions.

Diffraction A propagation phenomenon that allows radio waves to propagate beyond obstructions via secondary waves created by the obstruction and without the LoS between the transmitter and receiver. Classic types of diffractions are smooth Earth and knife-edge.

Dish antenna A dish-like antenna used to link communications sites by radio transmission of voice or data. Also called microwave antenna, parabolic antenna, or microwave dish antenna.

Diversity It is a technique to reduce the effects of radio fading. Types include space, frequency, angle, and hybrid.

DoD U.S. Department of Defense

DOE U.S. Department of Energy

DOT U.S. Department of Transportation.

Drop-insert A functionality provided in analog and digital repeaters, where radio-system specific control and service channels and possibly part of the payload is made available for local traffic and system management and maintenance.

DS0 Digital signal, level 0; 64 kbps (level zero [0]).

DS1 Digital signal, level 1; 1.544 Mbps, the North American standard.

DS3 Digital signal, level 3; 44.736 Mbps, the North American standard.

DSL Digital subscriber line. *See also* xDSL.

DS/SS Direct sequence spread spectrum.

DSU Digital service unit.

DSX-1 Digital service cross-connect, level 1; part of the DS1 specification.

DTE Data terminal equipment.

DTED Digital terrain elevation data.

Duct A channel for holding and protecting conductors and cables, made of metal or an insulating material (PVC, for example), usually circular in cross section like a pipe; sometimes also referred to as conduit.

Ducting Ducting is guided propagation of radio waves inside a tropospheric radio duct. For microwave links this type of anomalous radio propagation mechanism can be a serious problem.

Duplexer A duplexer is a device that allows bidirectional (duplex) communication over a single path. In radar and radio communications systems, it isolates the receiver from the transmitter while allowing them to share a common antenna. Most radio systems include a duplexer.

DUT Device under test.

DWDM Dense wavelength division multiplexing.

DXC Digital cross-connect.

E1 The European equivalent of a North-American T1. Equals 2.048 Mbps.

EC European Commission. The European Commission embodies and upholds the general interests of the European Union and is the driving force in the Union's institutional system.

ECC Elliptic curve cryptography

EEE Energy efficient Ethernet.

EFS Error-free seconds. Seconds without any errors.

EIA Electronic Industries Alliance (formerly Electronic Industries Association). It specifies electrical transmission standards, including those used in networking.

EIRP Effective isotropic radiated power in decibels per watt or decibels per meter.

EM Element manager; it is the radio link maintenance and control software provided with the radio terminal.

EMC Electromagnetic compatibility. EMC is the study of the unintentional generation, propagation, and reception of electromagnetic energy and the unwanted effects [electromagnetic interference (EMI)] that such energy may induce to equipment in the close proximity of each other.

EMS Energy management systems. A device in the customer's premises (hardware and software) designed to control the operation of other energy devices according to customer preferences and objectives such as reducing energy costs or maintaining comfort or convenience. Controlled devices could include, but are not limited to, thermostats, lighting, and smart appliances. Among other control inputs, an energy management device can accept energy-pricing signals from a utility or third-party energy services provider.

EPL Ethernet private line. Replaces a TDM private line.

EPLAN Ethernet private LAN. EPL is a data service defined by the Metro Ethernet Forum, providing a point-to-point Ethernet connection between a pair of dedicated user-network interfaces (UNIs), with a high degree of transparency.

EPSR Ethernet protection switching ring

ERC European Radiocommunications Committee. The ERC was established by CEPT to develop radio communications policy and to coordinate frequency, regulatory, and technical matters concerning radio communications.

ERP Effective radiated power (in a given direction). The product (or sum, if transmit power and the gain are given in logarithmic units) of the power supplied to the antenna and its gain relative to a half-wave dipole in a given direction.

ETSI European Telecommunications Standards Institute. A body formed by the European Commission in 1988. ETSI, which includes vendors and operators, aims to define standards that will enable the European market for telecommunications to function as a single market.

EVC Ethernet virtual circuit (connection).EVC is a logical relationship between Ethernet UNIs in a provider-based Ethernet service.

EVPL Ethernet virtual private line. Replaces frame relay or ATM services.

FAA Federal Aviation Administration.

Fade margin A design allowance that provides for sufficient system gain or sensitivity to accommodate expected fading, for the purpose of ensuring that the required quality of service is maintained. In radio communications systems, the difference in decibels between the power level at the receiver under nonfading conditions, and the receiver threshold.

Fading The variation in signal strength from normal value. Fading is normally negative and can be either fast or slow. It is usually characterized by the distribution of fades, Gaussian, Rician, or Rayleigh.

FAT Factory acceptance test. Factory validation procedures and sometimes witnessed by the customer at the factory.

FCC Federal Communications Commission. Regulates interstate communications via such means as licenses, rates, tariffs, standards, and limitations. Commissioners are appointed by the U.S. president. In Canada, the same function is performed by Industry Canada.

FDD Frequency division duplex. FDD means that the transmitter and receiver operate at different carrier frequencies.

FDM Frequency division multiplexing. FDM means that the total bandwidth available to the system is divided into a series of nonoverlapping frequency subbands that are then assigned to each communications channel.

FDMA Frequency division multiple access.

FEC Forward error correction. An encoding technique that allows a limited number of errors in a digital stream to be corrected based on knowledge of the encoding scheme used.

FERC Federal Energy Regulatory Commission.

FHSS Frequency-hopping spread spectrum. A method of transmitting radio signals by rapidly switching a carrier among many frequency channels, using a pseudorandom sequence known to both transmitter and receiver. It is a type of a spread-spectrum system.

FOTS Fiber-optic transmission system.

Frame relay Protocol for packet-switched data communications.

FRN FCC Registration Number. This is a 10-digit number that is assigned to an entity registering with the FCC. Every entity doing business with the FCC that wishes to obtain a license or certificate, register an antenna structure, or submit FCC Remittance Advice Form 159 must obtain an FRN.

Free-Space Loss Free-space path loss is the signal energy lost in traversing a path in free space only, with no other obstructions or propagation issues.

Frequency response Attenuation versus frequency or, in other words, a plot of how a circuit or device responds to different frequencies.

Fresnel zones Areas of constructive and destructive interference created when an electromagnetic wave propagating in free space is reflected (multipath) or diffracted as the wave intersects obstacles. Diffraction is caused by the phenomenon of an obstacle in the path of an electromagnetic wave not casting a completely sharp shadow. Instead, some of the energy in the wave front is bent away from the LoS path into the shadowed area behind the obstacle.

FRU Field replaceable unit.

FSO Free-space optical. FSO is an optical communications technology that uses light propagating in free space to transmit data between two points. The optical links are usually implemented using infrared laser light.
FTP File transfer protocol.
Gain A measure of directivity. It is defined as the ratio of the radiation intensity in a given direction to the radiation intensity that would be obtained if the power accepted by the antenna were radiated equally in all directions (isotropically). Antenna gain is expressed in decibels relative to an isotropic radiator.
GE Gigabit Ethernet
Geodetic datum A reference that describes the position, orientation, and scale relationships of a reference ellipsoid to the Earth.
GIS Geographic information system. A common name for software tools to acquire and maintain digital geographic data.
GMT Greenwich mean time.
GND Ground (0V). In electrical engineering, ground or Earth is the reference point in an electrical circuit from which other voltages are measured.
GNSS Global navigation satellite systems. Standard generic term for satellite navigation systems that provide autonomous geospatial positioning with global coverage.
GOOSE Generic object-oriented substation event.
GoS Grade of service.
Grooming Consolidating or segregating traffic to achieve efficiency.
Guard band A set of frequencies or bandwidth used to prevent adjacent systems from interfering with each other.
HAN Home area network. A network that is used inside of a home.
HF High-frequency. Radio frequencies between 3 and 30 MHz.
HiperMAN WiMAX variation created by the European Telecommunications Standards Institute (ETSI) Broadband Radio Access Networks (BRAN) group. Operates in the 2–11-GHz range.
HMI Human machine interface.
Hybrid A number of RF circuits are referred to as hybrids. RF hybrids are based on transformers, and therefore RF splitters or RF combiners may be referred to as hybrids when they use transformer technology.
Hz Hertz; cycles per second. Unit of measurement for frequency.
IC Industry Canada.
IEC Interexchange carrier. It is defined as any carrier that provides inter-LATA communication, where a LATA is a local access and transport area. An IXC carries traffic, usually voice traffic, between telephone exchanges.
IED Intelligent electronic device. IED is a generic term used to describe any microprocessor-based equipment, aside from a computer.
IETF Internet Engineering Task Force. IETF is a large open international community of network designers, operators, vendors, and researchers concerned with the evolution of the Internet architecture and the smooth operation of the Internet.
IF Intermediate frequency. IF is a frequency to which a carrier frequency is shifted as an intermediate step in transmission or reception. The intermediate frequency is created by mixing the carrier signal with a local oscillator.
ILEC Incumbent local exchange carrier. The existing local exchange carrier in any given area, as opposed to CLECs.

IM interference *See* intermodulation interference.
IN Intelligent network. A capability in the public telecom network environment that allows new services. Also implies a well-developed network infrastructure.
INM Integrated network management.
Intermodulation interference Interference caused directly by the nonlinear mixing of two or more radio signals. It creates new carriers on frequencies that are sums and differences of harmonics of the original carriers. The mixing most commonly occurs inside the receiver or transmitter of a radio, creating a third signal that is radiated from the antenna out to other radios. The mixing also can occur outside a radio in the transmission line or through rusty tower bolts or guy wires (passive intermodulation).
IP Internet protocol. *See also* TCP/IP.
ISM Instructional, scientific, and medical; microwave bands (2.4-GHz and 5.8-GHz) that do not require licensing in the United States.
ISO International Organization for Standardization. ISO is responsible for a wide range of standards, including many that are relevant to networking (OSI reference model and the OSI protocol suite, for example).
Isotropic antenna An antenna that radiates in all directions (about a point) with a gain of unity (not a real antenna, but a useful concept in antenna theory). Used as a 0-dB gain reference in directivity calculation (gain).
ISTO Industry Standards and Technology Organization. A not-for-profit membership organization, affiliated with IEEE and IEEE Standards Association.
ITU International Telecommunications Union. Based in Geneva, the ITU is an organization of the United Nations that oversees telecommunications standards around the world.
ITU-R Recommendations The international technical standards developed by the radiocommunication sector (formerly CCIR) of the ITU.
Kbps Kilobits per second
LAN Local area network. A communications network covering a small physical area, like a home, office, or small group of buildings, such as a school, campus, or an airport.
Latency The time it takes for a packet to travel from the origin to the destination and back (or only one way in utility communications circuits). It excludes the processing time at the destination. Latency includes the media access time and any queuing and propagation delays.
LBO Line buildout (I/O cable distance compensator).
LCT Local craft terminal; a PC with a web browser used for setup and configuration of microwave (or any other) terminal hardware.
Leased line A permanent telephone connection between two points set up by a telecommunications common carrier. Typically, leased lines are used by businesses to connect geographically distant offices. The primary factors affecting the monthly fee are distance between end points and the bandwidth of the circuit.
LiDAR Light detection and ranging.
LoS Line-of-sight. An unobstructed radio path or link between the transmitting and receiving antennas of a communications system (i.e., the transmission path is not established by or dependent upon reflection or diffraction).
LTE Long-term evolution; evolved air interface based on OFDMA, and utilizing FDD or TDD. In another context it can also mean *line terminal equipment*.

LTS Laser transmission system. *See* FSO.

MAC Media access control. MAC is a sublayer of the data link layer specified in the seven-layer OSI model (layer 2).

MAN Metropolitan area network. A metropolitan area network (MAN) is a large computer network that spans a metropolitan area or campus. Its geographic scope falls between a WAN and LAN.

MAS Multiple address radio.

MDIX Medium dependent interface crossover; if the Ethernet connections implement auto-MDIX, either straight or crossover Ethernet cables can be used, independent of the wiring of the connected device.

MDM Meter data management.

MEF Metro Ethernet Forum. MEF was founded in 2001 as a nonprofit international industry consortium, dedicated to worldwide adoption of carrier Ethernet networks and services.

MGB Master ground bar.

MHSB Monitored hot standby. Redundant (1+1) radio configuration, used to increase reliability of the microwave link hardware.

MIB Management information base. A file required by SNMP to describe the information that can be accessed for each network device.

MIMO Multiple-input multiple-output; minimizes signal fading due to path obstructions or atmospheric disturbances.

Minimum bending radius The amount of bend that a fiber or copper cable can withstand before experiencing performance problems.

MPLS Multiprotocol label switching (formerly known as tag switching). MPLS is a mechanism in high-performance telecommunications networks that directs and carries data from one network node to the next. The essence of MPLS is the generation of a short fixed-length label as a representation of an IP packet's header. MPLS is a highly scalable, protocol agnostic, data-carrying mechanism. It is called "multiprotocol" because it works with different network protocols.

MPLS-TE MPLS traffic engineering (formerly known as "RRR" or resource reservation routing); the label switching used to improve traffic performance and efficient use of network resources.

MTBO Mean time between outages.

MTBF Mean time between failures.

MTR Mean time to restore.

MTTR Mean time to repair. The average time required to repair or replace a failed device.

MUX Multiplexer (multiplexing). Combining two or more signals into a single bit stream that can be individually recovered.

Narrowband A classification of the information capacity or bandwidth of a communications channel. Narrowband is generally taken to mean a bandwidth of 64 kbps or below.

NATE National Association of Tower Erectors. NATE is a nonprofit trade association providing a unified voice for tower erection, maintenance, and service companies.

NCRP National Council for Radiation Protection and Measurement. NCRP was chartered by the U.S. Congress in 1964 as the National Council on Radiation Protection and Measurements.

NEBS Network equipment building standards. Originally developed by Bell Telephone Laboratories in the 1970s and expanded by Bellcore. NEBS covers a large range of requirements including criteria for personnel safety, protection of property, and operational continuity. NEBS is a major test of quality that is valuable for any organization supplying or purchasing network equipment.

NEC National Electrical Code. The National Electrical Code (NEC) is a United States standard for the safe installation of electrical wiring and equipment. It is part of the National Fire Codes series (NFPA 70) published by the National Fire Protection Association (NFPA). The code is updated every three years.

It can also mean *Nippon Electric Company.*

NERC North American Electric Reliability Corporation. NERC is a not-for-profit international regulatory authority whose mission is to ensure the reliability of the bulk power system in North America. NERC's area of responsibility spans the continental United States, Canada, and the northern portion of Baja California, Mexico.

NEXT Near-end crosstalk.

NIST National Institute of Standards and Technology. Founded in 1901, NIST is a nonregulatory federal agency within the U.S. Department of Commerce.

NIU Network interface unit; test unit installed at the demarcation point.

NLOS Non-line-of-sight. Non-line-of-sight (NLOS) or near-line-of-sight is a term used to describe radio transmission across a path that is partially obstructed, usually by a physical object in the Fresnel zone.

NMS Network management system. NMS is a combination of hardware and software used to monitor and administer a network. Individual network elements (NEs) in a network are managed by an element management system.

NNI Network-to-network interface. Demarcation between carrier Ethernet networks operated by one or more carriers.

NOAA National Oceanic and Atmospheric Administration.

NOC Network operations center. Although the term NOC is normally used when referring to telecommunications providers, there are a growing number of other organizations such as public utilities (e.g., SCADA) and private companies that also have such centers, both to manage their internal networks and to provide monitoring services.

NTIA National Telecommunications and Information Administration. NTIA is an agency in the U.S. Department of Commerce that serves as the executive branch agency principally responsible for advising the president on telecommunications and information policies.NTIA also manages the federal use of spectrum.

OA&MP Operations, administration, and maintenance provisioning.

OC-n Optical carrier, level n. For example, OC-1 is an optical carrier level 1 (51.84 Mbps), OC-3 is an optical carrier level 3 (155 Mbps), and so on.

OEM Original equipment manufacturer. A company whose products are used as components in another company's product. The OEM will generally work closely with the company that sells the finished product and customize the designs based on their needs.

OFDM Orthogonal frequency division multiplexing.

Omnidirectional antenna An antenna that radiates and receives radio signals equally in all directions in azimuth.

OPGW Optical ground wire. In electrical transmission systems, a ground wire that includes optical fibers to provide a communications link.

OPEX Operating expenses.

OSHA Occupational Safety and Health Administration. Created in 1971, OSHA's mission is to prevent work-related injuries, illnesses, and deaths.

OSI Open system interconnection. A seven-layer architecture model for communications systems developed by ISO and used as a reference model for most network architectures today.

OTDR Optical time domain reflectometer. OTDR is an opto-electronic instrument used for measurements on an optical fiber.

Packet switching A method of handling high-volume traffic that allows for efficient sharing of network resources, as packets from different sources can all be sent over the same channel in the bit stream. A packet-switched network breaks up the information into digital packets that are addressed and individually routed and then reassembled in the correct sequence at the destination. These networks allow the medium to be shared, so they are more efficient than circuit-switched networks.

Packetized voice Any means by which voice traffic is split into packets and then transferred to its end destination. This category includes both IP and Internet telephony.

PAM Pulse amplitude modulation. PAM is a form of signal modulation where the message information is encoded in the amplitude of a series of signal pulses.

Path loss Amount of loss introduced by the propagation environment between a transmitter and receiver.

PBX Private branch exchange; a private telephone switching system.

PCM Pulse code modulation. A digital representation of an analog signal; the magnitude of the signal is sampled regularly at uniform intervals and then quantized to a series of symbols in a numeric (usually binary) code.

PDH Plesiochronous digital hierarchy. PDH is an older technology used in telecommunications networks to transport data over digital transport equipment such as fiber-optic and microwave radio systems. PDH was mostly replaced by synchronous digital hierarchy (SDH) or synchronous optical networking (SONET) equipment in most telecommunications networks.

PDV Packet delay variation.

PHY Physical layer. The physical layer consists of the basic hardware transmission technologies of a network (the first and lowest layer in the seven-layer OSI model of computer networking).

PL Private line; a leased line, not switched.

PLC Power line carrier. PLC carries data on a conductor that is also used simultaneously for AC electric power transmission or electric power distribution to consumers.

Plenum An air-handling space such as that found above drop-ceiling tiles or in raised floors. Also, a fire-code rating for an indoor cable.

PMD Polarization mode dispersion.

PMU Phasor measurement units. Monitors that take observations at high speed (typically 30 observations per second—compared to one every 4 seconds using conventional technology). Each measurement is time-stamped according to a common time reference. Time stamping allows synchrophasors from different utilities to be time-aligned (or "synchronized") and combined together providing a precise and comprehensive view of the entire interconnection.

PoE Power over Ethernet. PoE is a technology that enables the safe transfer of DC electrical power along with data over the same standard network cabling (Cat 5/6).

POI Point of interface.

Point-to-multipoint system A radio system with a central master station; it communicates with a network of remote stations for data gathering and control. Sometimes called a *multiple address system* (MAS).

POP Point of presence. In telecommunications, a point-of-presence (POP) is a demarcation point or interface point between different communications systems.

POTS Plain old telephone service. A two-wire twisted pair copper circuits, providing bidirectional, full-duplex, voice band communications services. Powered by -48 VDC sent by the local central office equipment. Long-distance lines separate transmit and receive audio paths and use four-wire cable (two pairs).

PPP Point-to-point protocol. In networking, PPP is a data link protocol commonly used to establish a direct connection between two networking nodes. It can provide connection authentication, transmission encryption privacy, and compression. PPP is used over many types of physical media.

Precipitation rate A measure of the intensity of precipitation—generally expressed in millimeters (or inches) per hour. Also called *rainfall rate* or *rain rate*.

Protection relay A relay designed to initiate disconnection of a part of an electrical installation or to operate a warning signal, in the case of a fault or other abnormal condition in the installation.

PRS Primary reference source. The master clocking source in a network.

PSTN Public switched telephone network. The traditional, wired telephone network.

PTT Post telephone and telegraph company. A governmental agency in many countries.

PUC Public Utilities Commission. A state regulatory body.

PSK Phase-shift keying. A digital modulation scheme that conveys information by changing, or modulating, the phase of a carrier wave.

QA Quality assurance. It refers to planned and systematic production processes that provide confidence in a product's suitability for its intended purpose.

QAM Quadrature amplitude modulation. In QAM, the amplitude of two waves, 90 degrees out-of-phase with each other (in quadrature) are changed (modulated or keyed) to represent the information signal. Amplitude-modulating two carriers in quadrature can be equivalently viewed as both amplitude-modulating and phase-modulating a single carrier.

QoE Quality of (user) experience; it is the overall acceptability of an application or service, (mostly audio and/or video), as perceived subjectively by the end user.

QoS Quality of service. A term used to characterize network availability, quality, and reliability.

QPSK Quadrature phase-shift keying. PSK is a digital modulation scheme that conveys information by changing, or modulating, the phase of a carrier wave. QPSK is sometimes known as quaternary or quadriphase PSK, 4-PSK, or 4-QAM. It uses four points on the constellation diagram. With four phases, QPSK can encode two bits per symbol,

QRSS Quasi-random signal sequence.

Radome A radio frequency transparent cover used to protect the antenna from wind load, snow and ice, or dust buildup that would otherwise cause excessive mechanical stress on the antenna and the tower structure and undesirable radiation pattern distortion. Add-on radomes are most often seen on parabolic antennas, but other antennas (including Yagi-Uda) could have it too.

Radiation pattern A graphical representation in either polar or rectangular coordinates of the spatial energy distribution of an antenna.

RBER Residual bit-error rate.

Receiver sensitivity The minimum RF (or optical) signal power level required at the input of a receiver for certain performance (e.g., BER).

Refraction A change in direction of propagating radio energy caused by a change in the refractive index, or density, of a medium.

Repeater A device used to regenerate an optical or electrical signal (of any frequency) to allow an increase in the system length.

RF Radio frequency.

RFI Request for information. Request sent to equipment suppliers and/or service providers to provide initial information for their hardware and/or services. Also, radio frequency interference.

RFID Radio frequency identification. RFID is used for identification and tracking using radio waves.

RFP Request for proposal. Request sent to equipment suppliers and/or service providers to provide complete and detailed solution (including pricing) for the specific item(s) that could be hardware and/or services.

RFQ Request for quotation, sent to equipment suppliers and/or service providers to provide price for the specific item that could be hardware and/or services.

Rigger Rigger (also called climber) is the member of the radio installation team responsible for installing the antenna and cabling on the transmission tower.

RJ-11 Standard four-wire connectors for phone lines.

RJ-45 IEEE 802.3 1Base-T networks.

RJ-48 Standard eight-wire connectors for T1 and ISDN termination and local area data channels/subrate digital services. Only four wires are usually used.

ROI Return on investment. The time it takes to pay back (or recoup) the money invested in a technology or strategy.

Rollout Implementing (deploying) a product or a system. Synonym of "buildout."

RPL Ring protection link.

RSL Received signal level. The signal level at the receiver input (from the antenna), usually expressed in dBm (decibels over 1 mW).

RSSI Received signal strength indicator. RSSI is a measurement of the power present in a received radio signal. RSSI output is often a DC analog level.

RSTP Rapid spanning tree. RSTP is an evolution of the STP, a link layer network protocol that ensures a loop-free topology for any bridged LAN. RSTP was created to provide faster recovery (convergence time) from topology changes.

RTU Remote terminal unit. Used for monitoring and reporting to a remote location (over a telephone line) the condition of an associated piece of equipment at predetermined times, when polled and when the equipment condition dictates.

RU Rack unit. One standard EIA rack unit equals 1.75 inch (44.5 mm).

Safety code 6 Series of safety codes prepared by the Consumer and Clinical Radiation Protection Bureau, Health Canada. These codes specify the requirements for the safe use of, or exposure to, radiation emitting devices.
SAR Specific absorption rate.
SCADA Supervisory control and data acquisition; the monitor/control of pipelines, railroads, power utility networks, and such entities.
SDH Synchronous digital hierarchy.
SES Severely errored seconds. Seconds with 10^{-3} BER or worse.
SFP Small form pluggable.
SLA Service level agreement. The contract between a provider and a customer that sets the amount of bandwidth and QoS, among other things.
SLC Subscriber loop carrier. It refers to equipment providing central office-like telephone interface functionality.
Smart jack Network interface unit with cross-connecting and monitoring capabilities.
SMR Specialized mobile radio. A conventional two-way radio system, or trunked radio system, operated by a service in the 800- or 900-MHz bands. Some systems with advanced features are referred to as an enhanced specialized mobile radio (ESMR).
SNMP Simple network management protocol. It is a set of protocols for managing complex networks. SNMP works by sending messages, called protocol data units (PDUs), to different parts of a network. SNMP-compliant devices, called agents, store data about themselves in management information bases (MIBs) and return this data to the SNMP requesters.
SNR Signal-to-noise ratio (also, S/N).
SNTP The simple network time protocol. A protocol for synchronizing the clocks of computer systems or IEDs over packet-switched, variable-latency data networks. With SNTP, an IED can obtain clock time over an Ethernet network.
SONET Synchronous optical network.
Space diversity A diversity technique widely used in radio communications. It consists of two receiver antennas physically (spatially) separated to provide uncorrelated receiver signals.
Spread spectrum Spread-spectrum techniques are methods by which electromagnetic energy generated in a particular bandwidth is deliberately spread in the frequency domain, resulting in a signal with a wider bandwidth.
Splitter An RF splitter is the reverse of a combiner—in fact splitters and combiners utilize exactly the same circuits; the inputs for one form the outputs for the other. As the signal is split a number of ways, there is an associated reduction in signal level between the input and the output dependent upon the number of outputs for which the signal is shared.
SRDM Subrate digital multiplexer. Digital multiplexers, operating at the DS1 level, that are used to provide data channels at 64 kbps (DS0 level) and at various subrates such as 2.4, 4.8, 9.6, 19.2, and 56 kbps.
Standard radio atmosphere An atmosphere having the standard refractivity gradient.
STDM Statistical time division multiplexing. Time slots are assigned to signals dynamically to make better use of the bandwidth.

STM Synchronous transfer module. An ITU-defined communications method that transmits a group of time division multiplexed (TDM) streams synchronized to a common reference clock. It reserves bandwidth according to a rigid hierarchy, regardless of actual channel usage.
STM–N Synchronous transfer module—Level N. (STM-1 equals 155 Mbps.)
STP Spanning tree protocol. Defined by standard IEEE 802.1d, a scheme used by layer 2 switches to automatically inactivate certain network links so that traffic will have only one path between a specific source and destination and will not travel endlessly in loops. Spanning tree is a self-learning protocol that automatically reconfigures itself if any network link fails to send traffic over another path but this reconfiguration time can be relatively slow in light of today's networking speeds. *See also* rapid spanning tree protocol (RSTP).
Stratum Level of clock source used to categorize accuracy.
STS Synchronous transport signal.
Subrefraction Refraction for which the refractivity gradient is greater (i.e., positive or less negative) than the standard refractivity gradient.
Substation An area of equipment containing switches, circuit breakers, buses, and transformers for switching power circuits and to transform power from one voltage to another or from one system to another. A "switching station" is a station where transmission lines are connected without power transformers.
Super-refraction Refraction for which the refractivity gradient is less (i.e., more negative) than the standard refractivity gradient.
SWR Standing wave ratio.
Synchrophasors Precise grid measurements now available from monitors called PMUs. Synchrophasors enable a better indication of grid stress, and can be used to trigger corrective actions to maintain reliability.
Synchronous Type of transmission in which the transmission and reception of all data is synchronized by a common clock, and the data is usually transmitted in blocks rather than individual characters.
T1 A North-American transmission standard of 1.544 Mbps.
TBOS Telemetry byte oriented serial. TBOS is a communications protocol for the microcomputer link that monitors and controls remote equipment over an EIA-422 serial telemetry communications line. E-telemetry systems were developed as a method to monitor and control diverse network elements from a remote, centralized location.
TCP/IP Transmission control protocol/Internet protocol. The data protocols used for the Internet.
TDM Time division multiplexing. TDM is a type of digital multiplexing used for circuit-switched communication with a fixed number of channels and constant bandwidth per channel.
TDMA Time division multiple access.
TE Terminal equipment.
Telcordia Formerly Bellcore, it provides certain centralized research and standards coordination for the regional Bell operating companies (RBOC)s. It also coordinates security and emergency preparedness for the U.S. government. Bellcore was formed in 1984 when AT&T was broken up into the seven RBOCs. Today, it is part of Ericsson.

Telemetry A remotely accessible system (RTU) with local control and automation functionality.

Temperature inversion The phenomenon in which warm air is formed above a layer of cool air that is near the Earth's surface. This phenomenon very often causes abnormal radio propagation conditions.

TIA Telecommunications Industry Association. The U.S. telecom industry standards body.

TM Terminal multiplexer.

TMN Total/Telecommunications Management Network (ITU-T Recommendation M.3010).

TND Transmission (transport) network design.

UHF Ultra-high frequency. UHF designates a range of electromagnetic waves with frequencies between 300 MHz and 3 GHz.

UL Underwriters Laboratory. A nonprofit laboratory that examines and tests devices, materials, and systems for safety and that has begun to establish safety standards.

ULS Universal licensing system. Online system for radio licensing and research.

UNI User-to-network interface. Physical interface/demarcation between the service provider, cable operator, carrier, and subscriber.

U-NII Unlicensed National Information Infrastructure; a 5-GHz microwave band that does not require licensing in the United States.

UPS Uninterruptible power supply. Also known as a battery backup, provides emergency power from a separate source when utility power is not available, typically for up to 30 minutes until an auxiliary power supply can be turned on, utility power restored, or equipment safely shut down.

UPSR Unidirectional path-switched ring. A two-fiber configuration is a simple SONET protection mechanism. All payloads are carried clockwise on a single fiber. The payloads are duplicated counterclockwise on a second (protection) fiber, which is ignored in the absence of failure.

USGS United States Geological Survey.

UTC Utilities Telecom Council. Also, *Universal Time Coordinated*; an international time standard. This is the current term for what used to be known as Greenwich mean time (GMT).

UTM Universal Transverse Mercator. UTM coordinate system is a grid-based method of specifying locations on the surface of the Earth.

UTP Unshielded twisted pair.

UWB Ultra-wideband. UWB is a radio frequency band (not a technology or system) that can be used at very low energy levels for short-range/broadband communications.

V.24 Serial data communications interface. An ITU standard (1964) that defines the functions of all circuits for the EIA-232 interface but it does not describe the connectors or pin assignments; those are defined in ISO 2110. In the United States, EIA-232 incorporates the control signal definition of V.24, the electrical characteristics of V.28 and the connector and pin assignments defined in ISO 2110.

VF Voice frequency. A frequency within the audio range. In telephony, the usable voice frequency band ranges from approximately 300 to 3,400 Hz. The bandwidth allocated for a single voice-frequency transmission channel is usually 4 kHz, including guard bands.

VHF Very high frequency. VHF is the radio frequency range from 30 to 300 MHz.

VLAN Virtual LAN. VLANs can logically group networks and users regardless of their physical location.

VoIP Voice over IP. A general term for a family of transmission technologies for delivery of voice communications over IP networks such as the Internet or other packet-switched networks.

VPN Virtual private network.

VRLA Valve-regulated lead-acid battery.

VSWR Voltage standing wave ratio. The voltage standing wave ratio is a measure of how well a load is impedance-matched to a source. The value of VSWR is always expressed as a ratio with number one in the denominator (2:1, 3:1, 10: 1, etc.)

WAAS Wide area augmentation system. A system of satellites and ground stations that provide GPS signal corrections, giving better position accuracy, and currently only available in North America.

WAN Wide area network. A wide area network (WAN) is a computer network that covers a broad area and used to connect LANs and other types of networks together.

WDM Wavelength division multiplexing.

WECC Western Electricity Coordinating Council. WECC is the regional entity responsible for coordinating and promoting bulk electric system reliability in the Western Interconnection. WECC is geographically the largest and most diverse of the eight regional entities that have delegation agreements with the North American Electric Reliability Corporation (NERC). WECC's service territory extends from Canada to Mexico and includes the provinces of Alberta and British Columbia, the northern portion of Baja California, Mexico, and all or portions of the 14 Western states between.

Wideband A classification of the information capacity or bandwidth of a communications channel. Wideband is generally taken to mean a bandwidth between 64 kbps and 2 Mbps.

WiMAX Worldwide interoperability for microwave access. The current standard for broadband wireless MAN networks.

Wind load Loading imposed on an element (communications tower, for example) from the effects of the wind.

WLAN Wireless local area network.

ZOI Zone of influence.

About the Author

Dr. Harvey Lehpamer, P. Eng., has many years of experience in the planning, design, and deployment of wireless and wireline networks including microwave, fiber-optic and other communications systems in Europe, Canada, the United States, Africa, Mexico, and other parts of the world.

Lehpamer has worked for a number of well-known telecom and wireless, manufacturing, and power utility organizations around the world, either directly or as an outside consultant. He is a licensed professional engineer in the province of Ontario, Canada.

Today, Lehpamer works at POWER Engineers, Inc., in San Diego, California. He can be contacted by e-mail at HL_2@hotmail.com.

Index

L

M

N

Recent Artech House Titles in Power Engineering

Andres Carvallo, Series Editor

The Advanced Smart Grid: Edge Power Driving Sustainability, Second Edition, Andres Carvallo and John Cooper

Battery Management Systems, Volume I: Battery Modelings, Gregory L. Plett

Battery Management Systems for Large Lithium Ion Battery Packs, Davide Andrea

Battery Power Management for Portable Devices, Yevgen Barsukov and Jinrong Qian

Design and Analysis of Large Lithium-Ion Battery Systems, Shriram Santhanagopalan, Kandler Smith, Jeremy Neubauer, Gi-Heon Kim, Matthew Keyser, and Ahmad Pesaran

Designing Control Loops for Linear and Switching Power Supplies: A Tutorial Guide, Christophe Basso

Electric Systems Operations: Evolving to the Modern Grid, Mani Vadari

Energy Harvesting for Autonomous Systems, Stephen Beeby and Neil White

GIS for Enhanced Electric Utility Performance, Bill Meehan

Introduction to Power Electronics, Paul H. Chappell

Introduction to Power Utility Communications, Dr. Harvey Lehpamer

Battery Management Systems, Volume I, Battery Modelings, Gregory L. Plett

Power Line Communications in Practice, Xavier Carcelle

Power System State Estimation, Mukhtar Ahmad

A Systems Approach to Lithium-Ion Battery Management, Phil Weicker

Signal Processing for RF Circuit Impairment Mitigation in Wireless Communications, Xinping Huang, Zhiwen Zhu, and Henry Leung

Synergies for Sustainable Energy, Elvin Yüzügüllü

For further information on these and other Artech House titles,including previously considered out-of-print books now available through our In-Print-Forever® (IPF®) program, contact:

Artech House
685 Canton Street
Norwood, MA 02062
Phone: 781-769-9750
Fax: 781-769-6334
e-mail: artech@artechhouse.com

Artech House
16 Sussex Street
London SW1V 4RW UK
Phone: +44 (0)20 7596-8750
Fax: +44 (0)20 7630-0166
e-mail: artech-uk@artechhouse.com

Find us on the World Wide Web at: www.artechhouse.com